LABORATORY MANUAL
for
THE ART OF ELECTRONICS

LABORATORY MANUAL
for
THE ART OF ELECTRONICS

Paul Horowitz
Ian Robinson

Harvard University

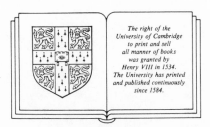

The right of the
University of Cambridge
to print and sell
all manner of books
was granted by
Henry VIII in 1534.
The University has printed
and published continuously
since 1584.

CAMBRIDGE UNIVERSITY PRESS
Cambridge
London New York
New Rochelle Melbourne Sydney

Published by the Press Syndicate of the University of Cambridge
The Pitt Building, Trumpington Street, Cambridge CB2 1RP
32 East 57th Street, New York, NY 10022, USA
10 Stamford Road, Oakleigh, Melbourne 3166, Australia

First published 1981
Reprinted 1983 (twice), 1984, 1986, 1987

Printed in the United States of America

Library of Congress Cataloging in Publication Data
Horowitz, Paul, 1942–

Laboratory manual for The art of electronics.

1. Electronics–Laboratory manuals.   2. Electronic
circuit design–Laboratory manuals.   I. Robinson,
Ian C.   II. Horowitz, Paul, 1942-   . Art of
electronics.   III. Title.
TK7818.H64   621.381   81-1413   AACR2
ISBN 0 521 24265 7 hard covers
ISBN 0 521 28510 0 paperback

CONTENTS

INTRODUCTION

This laboratory manual is intended to accompany **The Art of Electronics**, by Horowitz and Hill (Cambridge University Press, New York, 1980). It consists of 23 lab exercises (and reading assignments) that we use in our "Laboratory Electronics" course at Harvard University, each lab requiring 2-3 hours for the average student. Our students spend two afternoons each week in the lab; thus the set of labs is well suited to the typical 12-15 week college semester. The choice of subjects (and reading) reflects our view of a one-semester course based on the text, and therefore does not cover all areas included in that volume.

The actual lab exercises utilize commonly available general-purpose lab instrumentation and standard electronics parts, and therefore should be adaptable to an electronics lab anywhere the text is used. The following apparatus should be available at each experimental setup:

> Dual-trace triggered scope (e.g., Tektronix 455 or 932A)
> Wide-range function generator (e.g., IEC F41)
> General-purpose breadboard (e.g., AP Powerace 103†)
> Digital Multimeter (Keithley 169, Fluke 8010A)
> VOM (e.g., Simpson 260-6P)
> Variable regulated power supply, 0-20V (e.g., Lambda LL-902-OV)
> Logic probe, TTL/CMOS (e.g., Hewlett-Packard 545A)
> Resistance substitution box
> Part assortment (see Lab Supplies Appendix)
> Cables and clip leads
> Hand tools: Long-nosed pliers, small screwdriver, stripper

> († There is no commercial equivalent to the in-house breadboards we use; the inexpensive AP unit is close, but lacks an internal oscillator, uncommitted pots, pulse generators, BNC connectors, and binding posts.)

We have found that the choice of "bargain" instruments is, in the long run, false economy. Furthermore, students deserve to learn electronics with high quality laboratory equipment, maintained in good operating condition.

We have included an appendix containing the pinouts of all active devices (transistors, IC's) used in these lab exercises. With the inclusion of an appendix of Z-80 data, this lab manual is self contained, and requires no additional data sheets or data books. An appendix listing all small parts needed for these labs has also been included, to assist in maintaining a complete stock of lab supplies.

Each lab is intended as one afternoon's work; students generally disassemble their circuits at the end, freeing the breadboard for the next class group. The five microprocessor labs, however, constitute a single large project, and are left assembled between lab sessions. In these labs we have employed the versatile Z-80 processor, which, with its static registers, permits simplified single-stepping. We have confined ourselves to the 8085 compatible instruction subset and opcodes, however, for consistency with the text.

In keeping with the spirit of **The Art of Electronics**, we have kept the workbook format informal, and have avoided entirely the practice of requiring lab writeups in a fixed format. We've chosen a spiral binding so the book will lie flat on the table, instead of flopping around maddeningly; the large margins leave room for valuable scribblings.

We are indebted to Tom Hayes for his many insightful suggestions, and his painstaking proofreading of the manuscript.

Cambridge, Massachusetts
January, 1981

A Note on the Type

In keeping with the spirit of the electronic revolution, this book was prepared by the authors in camera-ready form, using a Z-80 computer running the Aox, Inc., MATE editor and the Scroll Systems, Inc., Retroscroller™. The manuscript was printed on a Diablo Systems, Inc., HyType II printer, using a Qume 'Theme 11pt' variable-pitch print wheel. The figures were hand drawn by the authors.

Reading: Chapter 1.1 - 1.11, pp 1-17.
                  Appendix A (don't worry if there are things
                  you don't understand)
                  Appendix C.

Problems: Problems in text.
                  Additional Exercises 1,2.

1-1. Ohm's Law.

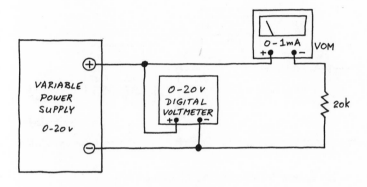

Figure 1.1. **V** vs **I** for a resistor.

Verify that Ohm's law is obeyed by measuring **V** and **I** for a
few voltages. Use a variable regulated dc supply, and the
hookup in figure 1.1. Note that voltages are measured
between points in the circuit, while currents are measured
through a part of a circuit. Therefore you usually have to
break the circuit to measure a current. After you've
measured a few values of **V** and **I** for the 20k resistor, try a
10k resistor instead.

   Interesting questions: The voltmeter is not measuring the
voltage at the place you want, namely across the resistor.
Does that matter? How can you fix the circuit so the
voltmeter measures what you want? When you've done that,
what about the accuracy of the current measurement? Can
you summarize by saying what an ideal voltmeter (or
ammeter) should do to the circuit under test? What does
that say about its "internal resistance"?

1-2. A Nonlinear Device.

Now perform the same measurement (**V** vs **I**) for a #47 lamp. Use the 100mA and 500mA scales on your VOM. <u>Do not exceed 6.5 volts!</u>

What is the "resistance" of the lamp? Does the question have any meaning?

1-3. The Diode.

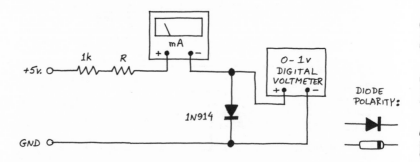

Figure 1.2. Diode **VI** measuring circuit.

Another non-linear device is the **diode.** Here, however, you can't just stick a voltage across it, as above. You'll see why after you've measured the diode's **V** vs **I**. Do that by wiring up the circuit in Figure 1.2.

Beginning with this exercise, use the prototyping "breadboard" at your bench -- have your instructor demonstrate which holes are connected to which, how to connect voltages and signals from the outside world, etc. In this circuit the 1k resistor limits the current to safe values. Vary **R** (use a 50k variable resistor ["pot"], a resistor substitution box, or a selection of various fixed resistors), and look at **V** vs **I**. Plot **V** vs $\log_{10}$**I** (keep the graph for use in lab 5). Now see what happens if you reverse the direction of the diode. How would you summarize the **V** vs **I** behavior of a diode?

Now explain what would happen if you were to put 5 volts across the diode (**Don't try it!**). Look at the diode data sheet in Appendix K to see what the manufacturer thinks would happen.

We'll do lots more with this important device; see, e.g., page 35ff in the text.

1-4. Voltage Divider.

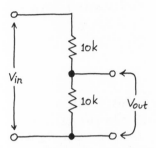

Figure 1.3 Voltage divider.

Construct the voltage divider in Exercise 1.9 (p. 10 of the text), reproduced here as figure 1.3. Apply V_{in} = 15 volts (use the dc voltages on the breadboard). Measure the (open circuit) output voltage. Then attach a 10k load and see what happens.

Now measure the short circuit current. From that, and the open circuit voltage, you can figure out the Thevenin equivalent circuit. Now build the Thevenin equivalent circuit, using the variable regulated dc supply as the voltage source, and check that its open circuit voltage and short circuit current are correct. Then attach a 10k load, just as you did with the original voltage divider, to see if it behaves identically.

1-5. Oscilloscope.

We'll be using the oscilloscope ("scope") frequently. Become familiar with its operation by generating a 1000 hertz (1kHz, 1000 cycles/sec) sine wave with the function generator and displaying it on the scope. Learn the operation of the scope's sweep and trigger controls. Try the triangle wave. Play with the vertical gain switch, the horizontal sweep speed selector, and the trigger controls, to learn what a misadjusted scope display looks like. Have your partner randomize some of the scope controls, then see if you can regain a sensible display (don't overdo it here!).

Switch the function generator to square waves and use the scope to measure the "risetime". What is coming out of the function generator's **SYNC OUT** connector? How about the **CALIBRATOR** on the scope? Put an "offset" onto the signal, if your function generator permits, then see what the **AC/DC** switch (located near the scope inputs) does. Try looking at pulses, say 1μs wide, 10kHz. Set the function generator to some frequency in the middle of its range, then try to make an accurate frequency measurement with the scope.

1-6. AC Voltage Divider.

First spend a minute thinking about this question: How would the analysis of the voltage divider be affected by an input voltage that changes with time (i.e., an input **signal**)? Now hook up the voltage divider from lab exercise 1-4, above, and see what it does to a 1kHz sine wave (use function generator and scope), comparing input and output signals. Explain in detail why it must act that way.

Reading: Chapter 1.12 - 1.20, pp 17-33; omit 'Power
                  in reactive circuits', pp 28-9.
                  Appendix B (if you need it).
                  Warning: This is by far the most
                  mathematical portion of the course. Don't
                  panic. Even if you don't understand the
                  math, you'll be able to understand the rest of
                  the book.

Problems: Problems in text.
                  Additional Exercises 3-6.

2-1. RC Circuit.

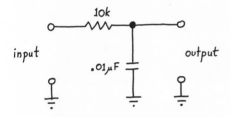

Figure 2.1. RC circuit.

Verify that the **RC** circuit behaves in the time domain as
described on pp 20-21. In particular, construct the circuit
in figure 2.1. Use a mylar capacitor (yellow tubular
package, with one lead sticking out each end). Drive the
circuit with a 500Hz square wave, and look at the output.
Measure the time constant by determining the time for the
output to drop to 37%. Does it equal the product **RC**? Try
varying the frequency of the square wave.

2-2. Differentiator.

Construct an **RC** differentiator (figure 2.2). Drive it with a
100kHz square wave, using the function generator with its
attenuator set to 20dB. Does the output make sense? Try a
100kHz triangle wave.

Interesting question: What is the impedance presented to
the signal generator by the circuit (assume no load at the
circuit's output) at f = 0? At infinite frequency? Questions

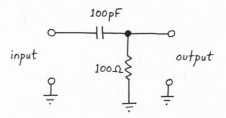

Figure 2.2. **RC** differentiator.

like this become important when the signal source is less
ideal than the function generators you are using.

2-3. Integrator.

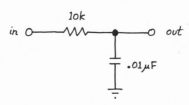

Figure 2.3. **RC** integrator.

Construct an integrator (figure 2.3). Drive it with a 100kHz
square wave at maximum output level (attenuator set at
0dB).   What is the input impedance at dc?   At infinite
frequency? Drive it with a triangle wave; what is the output
waveform called?

　　Note: Both the differentiator and integrator circuits are
approximations.   In Chapter 3 we will see how to make
"perfect" differentiators and integrators.

2-4. Low-pass Filter.

Construct the low-pass filter in figure 2.4. Where do you
calculate the -3dB point to be? Drive it with a sine wave,
sweeping over a large frequency range, to observe its low-
pass property; the 1kHz and 10kHz ranges should be most
useful.

　　Check to see if the filter attenuates 6dB/octave for
frequencies well above the -3dB point; in particular, measure
the output at 10 and 20 times f_{3dB}. While you're at it, look

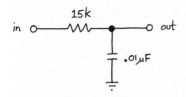

Figure 2.4. **RC** low-pass filter.

at the behavior of the phase shift vs frequency: What is the phase shift for $f \ll f_{3dB}$, $f = f_{3dB}$, and $f \gg f_{3dB}$?   Finally, measure the attenuation at $f = 2f_{3dB}$, and write down the attenuation figures at $f = 2f_{3dB}$ and $f = 10f_{3dB}$ for later use (section 2.9, below.).

2-5. High-pass Filter.

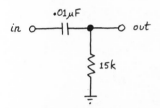

Figure 2.5. **RC** high-pass filter.

Construct a high-pass filter with the same components (figure 2.5).   Where is the -3dB point?   Check out its operation with sine waves; measure the frequency at which it attenuates by 3dB (70.7% of full amplitude).

Check to see if the output amplitude at low frequencies (well below the -3dB point) is proportional to frequency. What is the limiting phase shift, both at very low frequencies and at very high frequencies?

2-6. Filter Example I.

Figure 2.6 shows a way to see the "garbage" on the 110-volt power line.   First look at the output of the transformer, at **A.**   It should look more or less like a classical sine wave. (The transformer, incidentally, serves two purposes -- it reduces the 110Vac to a more reasonable 6.3V, and it "isolates" the circuit we're working on from the potentially lethal power line voltage)

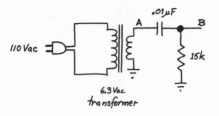

Figure 2.6. High-pass filter applied to the 60Hz ac power.

To see glitches and wiggles, look at **B**, the output of the high-pass filter. All kinds of interesting stuff should appear, some of it curiously time-dependent. What is the filter's attenuation at 60Hz (no complex arithmetic necessary)?

2-7. Filter Example II.

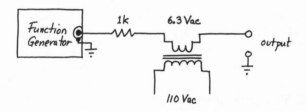

Figure 2.7. Composite signal, consisting of two sine waves. The 1k resistor protects the function generator in case the composite output is accidentally shorted to ground.

Try using high- and low-pass filters to clean up a composite signal. Make the signal by adding some 60Hz sine wave to the function generator output, as shown figure 2.7. Now run

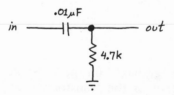

Figure 2.8. High-pass filter.

it through the high-pass filter of figure 2.8. Where is the -3dB point? Look at the resulting signal.

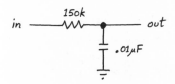

Figure 2.9. Low-pass filter.

Now run the composite signal through the low-pass filter of figure 2.9.   Where is the -3dB point?   Look at the resulting signal.   Why were the -3dB frequencies chosen where they were?

2-8. Blocking Capacitor.

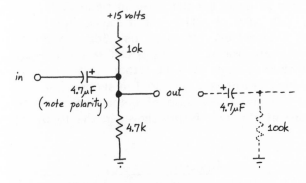

Figure 2.10. Blocking capacitor.

One of the most frequent applications of capacitors is for "blocking" dc while coupling an ac signal. You can think of it as a high-pass filter, with all signals of interest well above the -3dB point. To learn about blocking capacitors, wire up the circuitry on the left side of figure 2.10.

Drive it with the function generator, and look at the output on the scope, dc coupled. The circuit lets the ac signal "ride" on +5 volts. Next, add the dotted circuitry (another blocking capacitor), and observe the signal back at ground. What is the low frequency limit for this blocking circuit?

2-9. LC Filters.

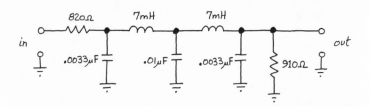

Figure 2.11. 5-pole Butterworth low-pass filter, designed using the procedure of Appendix H.

It is possible to construct filters (high pass, low pass, etc.) with a frequency response that is far more abrupt than the simple **RC** filters you have just built, by combining inductors with capacitors, or by using amplifiers (the latter are called "active filters" -- see Chapter 4 in the text). To get a taste of what can be done, try the filter in figure 2.11. It should have a -3dB point of 33kHz, and it should drop like a rock at higher frequencies. Measure its -3dB frequency, then measure its response at $f = 2f_{3dB}$ and at $f = 10f_{3dB}$. Compare these measurements with the rather "soft" response of the **RC** low-pass filter you measured in section 2.4, and the calculated response (i.e., ratio of output amplitude to input amplitude) of the two filters shown in the following table.

		frequency		
	0	f_{3dB}	$2f_{3dB}$	$10f_{3dB}$
RC	1.0	0.71	0.45	0.10
5-pole	1.0	0.71	0.031	0.00001

Reading: Finish Chapter 1, including "Power in reactive
                  circuits" (pp 28-29)
                  Appendix E.

Problems: Problems in text.
                  Additional Exercises 7,8.

3-1. LC Resonant Circuit.

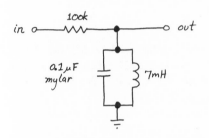

Figure 3.1. **LC** parallel tuned circuit.

Construct the parallel resonant circuit of figure 3.1.   Drive
it with sine waves (use the 10kHz range of the function
generator), and look at the response.   From the component
values calculate the resonant frequency; compare with the
measured value. As a virtuoso exercise, you might try using
the function generator in the **SWEEP** mode (if it has sweep
capability), with the scope's horizontal deflection driven by
the sweep ramp, to give a dynamic display of response vs
frequency. If you are successful at this, try to explain why
the displayed response grows funny wiggles as you increase
the sweep rate.

An even more interesting exercise involves using the
circuit as a "Fourier analyzer" -- its response is a measure
of the amount of 6kHz (approx.) present in the input
waveform. Try driving the circuit with a <u>square</u> wave at the
resonant frequency; note the amplitude of the (sine wave)
response. Now gradually lower the driving frequency until
you get another substantial response (it should occur for a
square wave at 1/3 the resonant frequency), and check the
response (it should be 1/3 as much). With some care you can
verify the amplitude and frequency of the first 5 or 6 terms
of the Fourier series. Can you think of a way to "measure
pi" with this circuit?

3-2. Half-wave Rectifier.

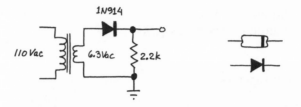

Figure 3.2. Half-wave rectifier.

Construct a half-wave rectifier circuit with a 6.3Vac (rms) transformer and a 1N914 diode (figure 3.2). Connect a 2.2k load, and look at the output on the scope. Is it what you expect? Polarity? Why is $V_{peak} > 6.3V$?

3-3. Full-wave Bridge Rectifier.

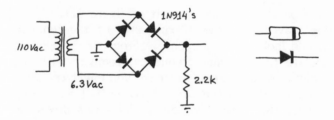

Figure 3.3. Full-wave bridge.

Now construct a full-wave bridge circuit (Fig 3.3). Be careful about polarities -- the band on the diode indicates cathode, as in the figure. Look at the output waveform (but **don't** attempt to look at the input -- the signal across the transformer's secondary -- with the scope's other channel at the same time; this would require connecting the second "ground" lead of the scope to one side of the secondary. What disaster would that cause?). Does it make sense? Why is the peak amplitude less than in the last circuit? How much should it be?   What would happen if you were to reverse any one of the four diodes? (**DON'T TRY IT!**).

Look at the region of the output waveform that is near zero volts. Why are there flat regions? Measure their duration, and explain.

3-4. Ripple.

Now connect a 15uF filter capacitor across the output (**IMPORTANT** - observe polarity). Does the output make sense? Calculate what the "ripple" amplitude should be, then measure it. Does it agree? (If not, have you assumed the wrong discharge time, by a factor of 2?)

Now put a 500μF capacitor across the output (again, be careful about polarity), and see if the ripple is reduced to the value you predict. This circuit is now a respectable voltage source, for loads of low current. To make a "power supply" of higher current capability, you'd use heftier diodes (e.g., 1N4002) and a larger capacitor.

3-5. Signal Diodes.

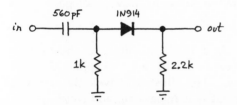

Figure 3.4. Rectified differentiator.

Use a diode to make a rectified differentiator (figure 3.4). Drive it with a square wave at 10kHz or so, at the function generator's maximum output amplitude. Look at the input and output, using both scope channels. Does it make sense? What does the 2.2k load resistor do? Try removing it.

3-6. Diode Clamp.

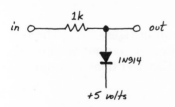

Figure 3.5. Diode clamp.

Construct the simple diode clamp circuit in figure 3.5. Drive it with a sine wave from your function generator, at maximum output amplitude, and observe the output.

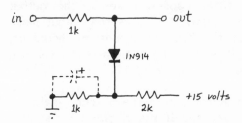

Figure 3.6. Clamp with voltage divider reference.

Now try using a voltage divider as the clamping voltage (figure 3.6). Drive it with a large sine wave, and examine the peak of the output waveform. Why is it rounded so much? (Hint: What is the impedance of the "voltage source" provided by the voltage divider?) To check your explanation, drive it with a triangle wave; compare with figure 1.83 in the text.

As a remedy, try adding a 15μF capacitor, as shown with dotted lines (note polarity). Try it out. Explain to your satisfaction why it works. This illustrates well the concept of a bypass capacitor. What is it bypassing, and why?

3-7. Diode Limiter.

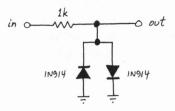

Figure 3.7. Diode limiter.

Build the simple diode limiter shown in figure 3.7. Drive it with sines, triangles, and square waves of various amplitudes. Describe what it does, and why. Can you think of a use for it?

3-8. Impedances of Test Instruments.

We mentioned in the first lab that measuring instruments (voltmeters, ammeters) should ideally leave the measured circuit unaffected. For instance, this implies an infinite impedance for voltmeters, and zero impedance for ammeters. Likewise, an oscilloscope should present an infinite input impedance, while power supplies and function generators should be zero-impedance sources.

Begin by measuring the internal resistance of the VOM on its 10V dc range. You won't need anything more than a dc voltage and a resistor, if you're clever. Next try the same measurement on the 50V dc range. Make sense? (Most needle-type VOM's are marked with a phrase such as "20,000 ohms per volt" on their dc voltage ranges; can you explain?) For further enlightenment, see the Box on Multimeters (text pp 8-9).

Now use a similar trick to measure the input resistance of the scope. Remember that it should be pretty large, if it's a good voltage measuring-instrument. As a voltage source use a 100Hz sine wave, rather than a dc voltage as above.

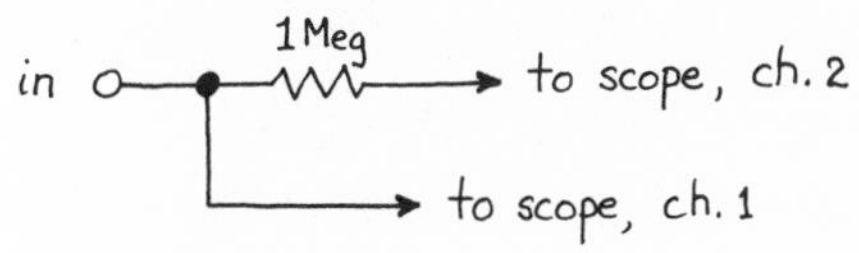

Figure 3.8. Circuit for measuring oscilloscope input impedance.

To measure the scope's input impedance, drive it with a signal in series with 1 megohm (figure 3.8). What is the low frequency (f < 1kHz) attenuation? Now raise the frequency. What happens? Explain, in terms of a model of the scope input as an R in parallel with a C. What are the approximate values of R and C? What remedy will make this circuit work as a divide-by-two signal attenuator at all frequencies? Try it!

Now go back and read the section entitled "Probes" in Appendix A (p. 641). Then get a 10x probe, and use it to look at the calibrator signal (usually a 1V, 1kHz square wave) available on the scope's front panel somewhere. Adjust the probe "compensation" screw to obtain a good square wave. Use 10x probes on your scope in all remaining lab exercises, like a professional!

Finally, measure the internal resistance of the function generator. **DON'T TRY TO DO IT WITH AN OHMMETER!** Instead, load the generator with a known resistor and watch the output drop. One value of R_{load} is enough to determine $R_{internal}$, but try several to see if you get a consistent value. Use a small signal, say 1 volt pp at 1kHz.

Reading: Chapter 2.01 - 2.08, pp 50-64.

Problems: Problems in text.
                Additional Exercises 1,3.
                Bad Circuits A,B,D,E,H,I.

4-1. Transistor Junctions Are Diodes.

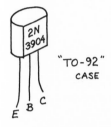

Figure 4.1. TO-92 transistor package.

Get a 2N3904 NPN transistor, identify its leads, and verify
that it looks like the object in figure 2.2 of the text by
measuring the BC and BE junctions with the VOM in the
Rx100 resistance range (don't use **Rx1** -- too much
current). What does figure 2.2 of the text suggest you
would find if you made a resistance measurement of the CE
junction? Try measuring it with the VOM. Remember this as
a method for spot checking a suspected bad transistor; it
must at least behave like a pair of diodes.

4-2. Transistor Current Gain.

Measure h_{FE} at several values of I_C with the circuit of figure
4.2. The 4.7k and 1k resistors limit the currents. Which
currents do they limit, and to what amounts?

     Try various values for **R**, using a resistor "substitution
box", e.g., 4.7Meg, 1Meg, 470k, 100k, 47k. Estimate the
base current in each case (V_{BE} = 0.6V, approximately) and,
from the measured collector current, calculate the beta
(h_{FE}).

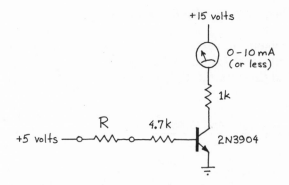

Figure 4.2. h$_{FE}$ measurement circuit.

4-3. Transistor Switch.

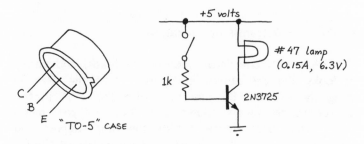

Figure 4.3. Transistor switch.

Try the circuit of Fig 4.3, in which the transistor operates as a switch, i.e., either saturated or off. Turn the base current on and off by pulling one end of the resistor out of the breadboard.

What is I$_B$, roughly? What is the minimum required beta?

Measure the saturation voltage, V$_{CE(sat)}$, with meter or scope.   Then parallel the base resistor with 150 ohms and note the improved V$_{CE(sat)}$. See Appendix G in the text for more on saturation.

4-4. Emitter Follower.

Wire up an NPN transistor as an emitter follower (figure 4.4).   Drive it with a sine wave that is symmetrical about zero volts (be sure the dc "offset" of the function generator is set to zero), and look with a scope at the poor replica that comes out.   Explain exactly why this happens.   Now try

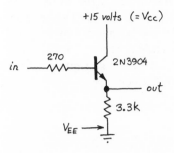

Figure 4.4. Emitter follower. The small base
resistor is often necessary to prevent oscillation.

connecting the emitter return (the point marked V_{EE}) to
-15V instead of ground, and look at the output. Explain.

4-5. Input and Output Impedance of Follower.

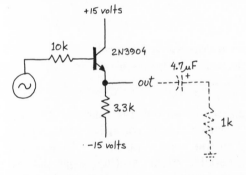

Figure 4.5. Circuit for measuring Z_{in} and Z_{out} of
the emitter follower.

In the last circuit replace the small base resistor with a 10k
resistor, in order to simulate a signal source of moderately
high impedance, i.e., low current capability (figure 4.5).

a) Measure Z_{out}, the output impedance of the follower, by
connecting a 1k load (with blocking capacitor -- why?) to
the output and observing the drop in output signal amplitude;
for this use a small input signal, less than a volt. (Viewing
the emitter follower's output as a signal source in series with
Z_{out} [Thevenin], the 1k load forms a divider at signal
frequencies, where the impedance of the blocking capacitor
is negligibly small. Take it from there.) Does the measured
value make sense?

b) Remove the 1k load. Now measure Z_{in}, the impedance looking into the transistor's base in this particular circuit configuration, by looking alternately at both sides of the 10k input resistor.   For this measurement the 3.3k emitter resistor is also the "load". Again, use a small signal. Does the result make sense (see p54 in the text)?

4-6. Single Supply Follower.

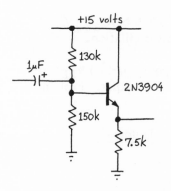

Figure 4.6. Single supply follower. A 270 ohm resistor in series with the base may be necessary if the circuit exhibits oscillations.

Figure 4.6 shows a properly biased emitter follower circuit, operating from a single positive supply voltage. This circuit comes from the example in the text on pp 57-8. Wire it up, and check it for the capability of generating large output swings before the onset of "clipping". For largest dynamic range, amplifier circuits should exhibit symmetrical clipping.

4-7. Current Source.

Construct the current source (sink) in figure 4.7. Slowly vary the 2.5k variable load, and look for changes in current measured by the VOM.   What happens at maximum resistance? Can you explain, in terms of voltage compliance of the current source?

Even within the compliance range, there are detectable variations in output current as the load is varied.   What causes these variations? Can you verify your explanation, by making appropriate measurements? (Hint: Two important assumptions were made in the initial explanation of the current source circuit on page 59 in the text.)

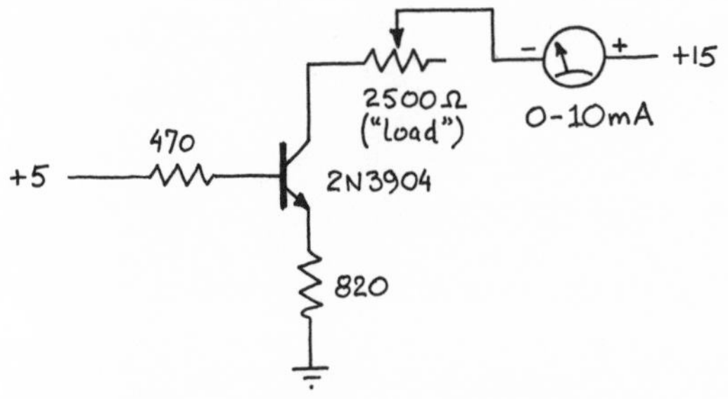

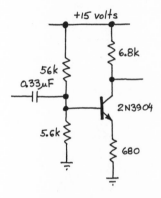

Figure 4.7. Transistor current source.

4-8. Common-emitter Amplifier.

Figure 4.8. Common-emitter amplifier.

Wire up the common emitter amplifier in figure 4.8. What should its voltage gain be? Check it out. Is the signal's phase inverted? Is the collector quiescent operating point (i.e., its resting voltage) right? How about the amplifier's low frequency -3dB point? What should the output impedance be? Check it by connecting a resistive load, with blocking capacitor.

4-9. Emitter Follower Buffer.

Hook an NPN emitter follower to the previous amplifier. Think carefully about coupling and bias. Use a 1k emitter resistor.

Measure output impedance again, using a small signal. Is the overall amplifier gain affected by the addition of the emitter follower?

Reading: Chapter 2.09 - 2.14, pp 64-77.

Problems: Problems in text.
                 Additional Exercises 2,7.
                 Bad Circuits C,F.

5-1. Dynamic Diode Curve Tracer.

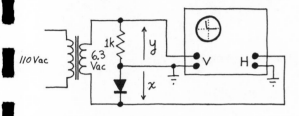

Figure 5.1. Dynamic curve tracer.

Wire up the **VI** curve tracer in the figure above. It uses the
horizontal and vertical scope inputs to provide a display of **I**
vs **V**. Explain how it works -- in particular, why can't you
use the function generator instead of the 6.3V transformer?

   Try it out. First use a 1N914. Be sure you know where
zero voltage and current are on the screen, by alternately
removing the input from V and H. Figure out the calibration
(mA/div, V/div), and then make a reasonably accurate plot
on graph paper. Compare it with the graph you made in lab
1-3. Stare at the diode display, to get a feeling for the
Ebers-Moll equation.   Then reverse the diode polarity.
Finally, replace the 1N914 (ordinary signal diode) by a 1N749
or equivalent (4.3V zener diode), and plot its characteristic
also.

5-2. Ebers-Moll Equation.

Wire up the circuit you used in Lab 4 to measure h_{FE} (figure
5.2). Again, use the substitution boxes for **R**, to generate
collector currents going from a few microamps to a few
milliamps. Plot the logarithmic increase of V_{BE} with I_C, and
confirm the "60mV/decade" law.

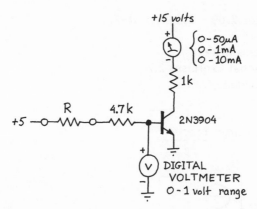

Figure 5.2. Circuit for measuring I_C vs V_{BE}.

5-3. Grounded Emitter Amplifier.

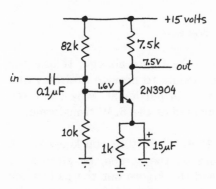

Figure 5.3. Grounded emitter amplifier.

Wire up the circuit in figure 5.3 (taken from the text, figure 2.35). First, check the quiescent collector voltage. Then drive it with a small <u>triangle</u> wave at 10kHz, at an amplitude that almost produces clipping (you'll need to use plenty of attenuation -- 40dB or more -- in the function generator). Does the output waveform look like Fig 5.4 (text figure 2.34)? Explain to yourself exactly why.

Now remove the 15µF capacitor, increase the drive amplitude (the gain is greatly reduced), and observe a full-swing triangle output without noticeable distortion. Measure the voltage gain -- does it agree with your prediction?

Restore the 15uF capacitor, and reduce the function generator output to the minimum possible. Predict the voltage gain at the quiescent point, using r_e. Measure it; does it agree?

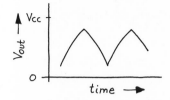

Figure 5.4. Large-swing output of grounded emitter
amplifier when driven by a triangle wave.

5-4. Biasing With DC Feedback.

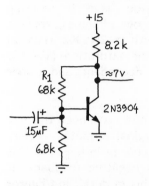

Figure 5.5. Grounded emitter amplifier with dc
feedback.

Wire up the grounded emitter amplifier with dc feedback
shown in figure 5.5 (text figure 2.40).   This arrangement
provides some measure of bias stability.   The nominal
collector quiescent point is $11V_{BE}$, or roughly 7 volts. If the
quiescent collector voltage were more than that, for
instance, the base divider would drive the transistor into
heavy conduction, restoring the proper operating point;
similarly, the proper operating point would be restored if the
quiescent point were to drop.

   Check to see if the quiescent collector voltage is
approximately correct. Since V_{BE} depends on temperature,
you should be able to shift the collector voltage a small
amount by warming the transistor between your fingers;
which way should it move?   In practice this slight
temperature sensitivity is not a major drawback; biasing a
grounded-emitter stage without such a feedback scheme is
considerably more uncertain, as we will now see.

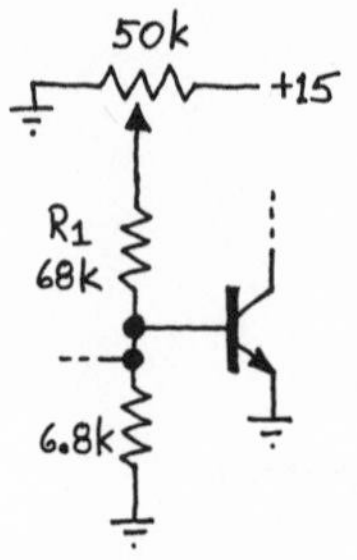

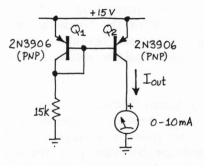

Figure 5.6. Poor biasing scheme for the grounded
emitter amplifier.

Now disconnect R_1 from the collector, and bias it instead
with a pot (figure 5.6).   After you have it working
satisfactorily (symmetrical swing without clipping), replace
the 2N3904 (typical beta of 100) with a 2N5962 (typical
beta of 1000), and note the collector saturation (V_C too
low). Such a bias scheme is very h_{FE}-dependent, and a poor
idea.

Leave the 2N5962 in the circuit, and reconnect the 68k
resistor (R_1) to the collector (the original circuit).   Verify
correct biasing, even with this large change in h_{FE}. In this
circuit the bias scheme also provides negative feedback at
signal frequencies. We'll talk about this in detail in Chapter
3; but to see that something interesting is happening, put a
6.8k resistor in series with the input signal, and note the
good linearity at large swing (use a triangle wave, again).

5-5. Current Mirror.

Figure 5.7. Classic PNP current mirror.

Build the classic current mirror (figure 5.7, same as text
figure 2.43). How closely does the output current equal the

programming current (you can calculate the latter without measuring anything)?   Now try squeezing one of the transistors with your fingers, to see how much I_{out} is affected by differences in temperature of Q_1Q_2. Does I_{out} increase or decrease? Explain.

Try substituting another 2N3906 for Q_2, to see what kind of spread of output current you are likely to get with unmatched transistors. Then put some resistance (5k to 10k) in series with the meter, to see if I_{out} varies with load voltage. How good a current source is the current mirror, compared with the current source you built in the last lab?

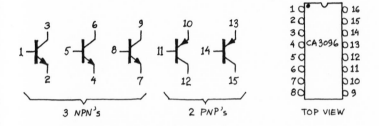

Figure 5.8. CA3096 pinout.

Now replace Q_1 and Q_2 by a monolithic transistor array, the CA3096. Figure 5.8 shows the "pinout". Measure I_{out}, using the matched PNP pair of the CA3096 in the original mirror circuit above. How closely does the output current match the programming current?   The best version of the CA3096 (the CA3096A) has V_{BE} matching of 0.15mV (typical), and 5mV (maximum); what ratios of collector current do these V_{BE} mismatches correspond to?

5-6. Crossover Distortion.

Explore "crossover distortion" by building the push-pull output stage in figure 5.9. Drive it with sinewaves of at least a few volts amplitude, in the neighborhood of 1kHz. Be sure the **OFFSET** control of the function generator is set to zero. Look closely at the output. (If things behave very strangely, you may have a "parasitic oscillation". It can be tamed by putting a 470 ohm resistor in series with the common base lead, and, if necessary, by adding a 100pF capacitor from the output to ground.)

Try running the amplitude up and down. Play with the dc offset control, if your function generator has one.

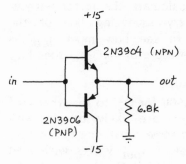

Figure 5.9. Complementary push-pull emitter follower.

Reading: Chapter 2.15 to end, pp 77-91.

Problems: Problems in text.
                Additional Exercises 4,5,6,8.
                Bad Circuit G.

6-1. Darlington.

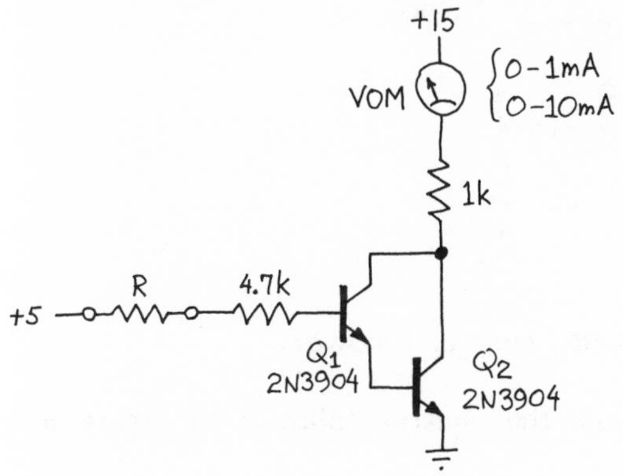

Figure 6.1. Darlington test circuit. Use a
substitution box for **R**.

Connect two 2N3904's in a Darlington, and measure its
characteristics with the circuit above.

a)   Begin by reducing **R** to a few thousand ohms, to bring
the transistor into good saturation; at this point $I_C = 15mA$.
Measure V_C (the "Darlington saturation voltage") and V_B of
Q_1 ("Darlington V_{BE}"). How do they compare with typical
single-transistor values?   (Measure Q_1 alone, if you don't
know, by grounding its emitter.) Explain.

b)   Increase **R** into the range of 500k-10Meg, in order to get
I_C down to a few milliamps.   Measure h_{FE} at collector
currents around 1mA and 10mA (as earlier, measure I_C, but
calculate I_B from R).

6-2. Superbeta.

Substitute a single superbeta 2N5962 for the Darlington pair,
and make the same set of measurements, namely V_{BE},

$V_{CE(sat)}$, and h_{FE} at collector currents around 1mA and 10mA. Does it meet the "typical" values graphed in figure 2.76 of the text?

6-3. Bootstrap.

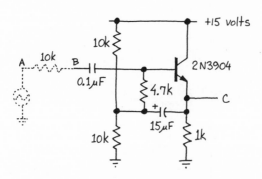

Figure 6.2. Bootstrapped emitter follower.

Begin by connecting up the emitter follower of figure 6.2 (text figure 2.63).

a) First, omit the 15µF capacitor. What should the input impedance be, approximately? Measure it, by connecting 10k in series with the function generator and noting the drop from A to B. Check that the output signal (at C) has the same amplitude as the signal at B (use an input frequency in the range of 10kHz-100kHz).

b) Now add the 15µF bootstrap capacitor. Again measure the input impedance by looking at both sides of the 10k series resistor with the scope. Make sure you understand where the improvement comes from.

c) Try removing the 10k series resistor and the bootstrap capacitor, then gradually increasing the drive amplitude, while looking at the output, until you have the function generator at full output level. What are those funny negative bumps at the output? (A 270 ohm resistor in series with the input should be added if you have oscillations.)

6-4. Differential Amplifier.

Wire up the circuit in figure 6.3. Predict what the quiescent collector voltages should be. Predict the differential and common-mode gains (don't forget about r_e).

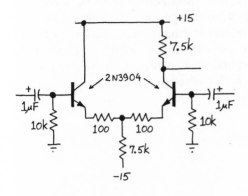

Figure 6.3. Differential amplifier.

Now measure those quantities: For G_{diff} ground one input and apply a small signal to the other; for G_{CM} tie both inputs together and drive with a moderate signal, say 1V pp.

6-5. Miller Effect.

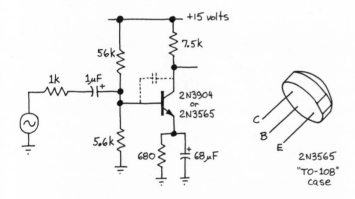

Figure 6.4. Grounded emitter NPN amplifier for exploring Miller effect.

Begin by constructing the high-gain (bypassed emitter) single-ended amplifier in figure 6.4.   Predict the voltage gain, then measure it (short the 1k input resistor).   Check that the collector quiescent voltage is reasonable.

Now restore the 1k series resistor (it simulates finite generator impedance, as you might have within a circuit). Measure the high-frequency 3dB point.

Now add a 33pF capacitor from collector to base (dotted lines). This swamps the transistor's junction capacitance of approximately 2pF, exaggerating the Miller effect. Remeasure the high-frequency 3dB point. Explain it quantitatively in terms of the effective capacitance to ground produced by the Miller effect.

LAB 7. OP AMPS I

Reading: Chapter 3.01 - 3.09, pp 92-102.

Problems: Problems in text.
                Bad Circuits B,D,E,F,G,I,K,M.

7-1. Op Amp Open-Loop Gain.

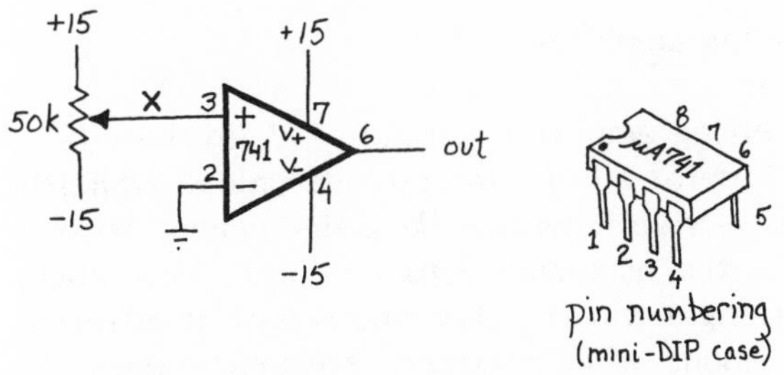

Figure 7.1. Open-loop test circuit.

Astound yourself by connecting up the circuit above, and watching the output voltage as you slowly twiddle the pot. Is the behavior consistent with the specifications, which claim "Gain(typical) = 200,000"?

Now break the input at the point marked **X**, tie the two inputs together, and insert a 1Meg resistor so you can measure the (small) input current by measuring the voltage drop across 1Meg with the digital multimeter (1 volt range). What is the measured input current? From the sign of the current deduce whether the input transistors are NPN or PNP; check by looking at the 741 schematic on page 104 of the text. Is the measured input current consistent with the specified value "$I_{bias} = 80nA(typ), 500nA(max)$"?

Although the input current of the 741 is quite small by ordinary transistor standards, and can often be ignored, you may need op-amps that approach the ideal (no input current) more closely. The popular 355 uses FET's (field-effect transistors) in the input stage for lower input current. Try one in place of the 741 -- they have the same "pinout". Is the specified input current of $I_{bias} = 0.03nA(typ), 0.2nA(max)$ measurable with your circuit?

7-2. Inverting Amplifier.

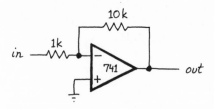

Figure 7.2. Inverting amplifier.

Construct the inverting amplifier in figure 7.2 (in these labs, use ±15 volt supplies for all op-amp circuits unless explicitly shown otherwise). Drive it with a 1kHz sine wave. What is the gain? What is the maximum output swing? How about linearity (try a triangle wave)? Try sine waves of different frequencies. At about what maximum frequency does the amplifier stop working well? Is this upper frequency limit amplitude-dependent?

Now go back to 1kHz sine waves. Measure the input impedance of this amplifier circuit by adding 1k in series with the input. Measure (or try to, anyway) the output impedance -- note that no blocking capacitor is needed (why?). Since the op-amp cannot supply more than several milliamps of output current, you will have to keep the signal quite small here. Note that a low value of "small-signal" output impedance doesn't necessarily mean that lots of power (output swing into a low value of load impedance) is available.

7-3. Non-inverting Amplifier.

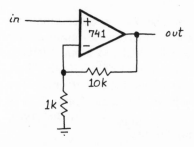

Figure 7.3. Non-inverting amplifier.

Wire up the non-inverting amplifier in Fig 7.3. What is the voltage gain (note -- it's not the same as the last circuit)?

Measure its input impedance, at 1kHz, by putting 100k, or 1Meg, in series with the input. Does this configuration maintain the low output impedance you measured for the inverting amplifier?

7-4. Follower.

Build a <u>follower</u> with a 741. Check out its performance; in particular, measure (if possible) Z_{in} and Z_{out}. How about its ability to follow a high-frequency signal? More on all these questions of op-amp limitations in the next lab.

7-5. Current Source.

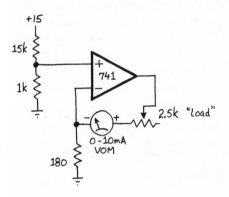

Figure 7.4. Current source.

Try an op-amp current source (figure 7.4). What should the current be? Vary the 2.5k pot and watch the current. If you can't see any variation, try using one of the digital multimeters.

Note that this current source, although far more precise and stable than our simple transistor current source, has the disadvantage of requiring a "floating" load (neither side connected to ground); in addition, it has significant speed limitations, in a situation where either the output current or load impedance varies at microsecond speeds.

7-6. Current-to-Voltage Converter.

Use an FPT100 phototransistor as a photodiode in the circuit of figure 7.5. Look at the output signal (if the dc level is more than 10 volts, reduce the feedback resistor to 1Meg).

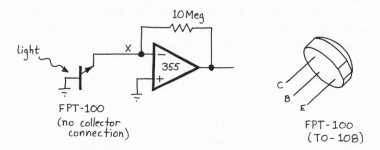

Figure 7.5. Photodiode photometer circuit.

What is the average dc output level, and what is the percentage "modulation" (the latter will be relatively large if the laboratory has fluorescent lights)?   What input photocurrent does the output level correspond to?   Try covering the phototransistor with your hand.   Look at the "summing junction" (point X) with the scope, as V_{out} varies. What should you see?

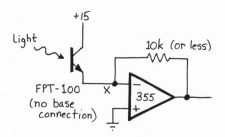

Figure 7.6. Phototransistor photometer circuit.

Now connect the FPT100 as a phototransistor (figure 7.6).   What is the average input photocurrent now?   What about the percentage modulation?   Look again at the summing junction.

7-7. Summing Amplifier.

The circuit in figure 7.7 is a BCD-to-analog converter (BCD means "binary coded decimal"), a weighted version of figure 3.18 in the text.   The 1-2-4-8 inputs are either +5V or 0V. The output is a voltage proportional to the value of the binary number at the input.

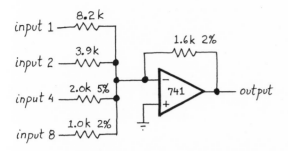

Figure 7.7. BCD weighted summing amplifier.

First figure out a) the polarity of the output voltage, and b) the conversion sensitivity (i.e., how many volts/unit). Then build it and try it out. What additional circuitry is needed in order to get a positive polarity output voltage?

7-8. Push-pull Buffer.

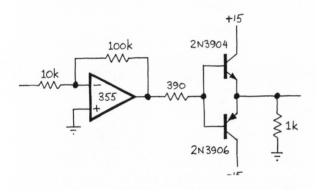

Figure 7.8. Inverting amplifier with push-pull buffer.

Build the circuit in figure 7.8. Drive it with a sine wave of 100Hz-500Hz. Look at the output of the op-amp, and then the output of the push-pull stage (make sure you have at least a few volts of output, and that the function generator is set for no dc offset). You should see classic crossover distortion.

Now reconnect the right side of the 100k resistor to the push-pull output (as in text figure 3.21), and once again look at the push-pull output. The crossover distortion should be eliminated now. If that is so, what should the signal at the output of the op-amp look like? Take a look.

LAB 8. OP AMPS II

Reading: Chapter 3.11 - 3.19, pp 102-122.

Problems: Problems in text.
Additional Exercises 1-4,6,7.
Bad Circuits A,C.

8-1. Op-amp Limitations.

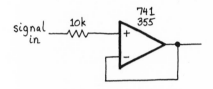

Figure 8.1. Slew rate measuring circuit. The series
resistor prevents damage if the input is driven beyond
the supply voltages.

1) **Slew Rate.** Begin by measuring slew rate with the
circuit above.

a) Drive the input with a square wave, in the neighborhood
of 1kHz, and look at the output with a scope. Measure the
slew rate by observing the slope of the transitions. See
what happens as the input amplitude is varied.

b) Switch to a sine wave, and measure the frequency at
which the output amplitude begins to drop, for an input level
of 10 volts (pp), say. Is this consistent with (a)?

Now go back and make the same pair of measurements
with the 741 replaced by a 355 (same pinouts). The 741
claims a "typical" slew rate of 0.5V/µs; the 355 claims
5V/µs. How do these values compare with your
measurements?

2) **Offset Voltage.** Now construct the x1000 non-
inverting amplifier in figure 8.2. Measure the offset voltage,
using the amplifier itself to amplify the input offset to
measurable levels. **Note**: Think carefully about what you
must do to the "in" terminal in order for your measurement
not to be confounded by the effects of input bias current,
which is typically about 0.08µA. Compare your measured
offset voltage with specs: V_{os} = 2mV(typ), 6mV(max).

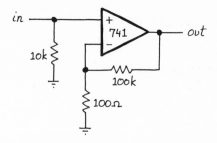

Figure 8.2. Offset measuring circuit.

Figure 8.3. 741 Offset trimming network.

3) **Offset Trim.** Trim the offset voltage to zero, using the recommended network (figure 8.3).

4) **Bias Current.** Now remove the connection from "in" to ground that you should have used in part (2), in order to measure the bias current. Explain how the 100k input resistor allows you to measure I_{bias}. Then compare your measurement with specs: $I_{bias} = 0.08\mu A$(typ), $0.5\mu A$(max).

5) **Offset Current.** Alter the circuit in such a way that both op-amp input terminals see 100k driving resistance, yet the overall voltage gain of the circuit is unchanged. This requires some thought (hint: you will need to add one resistor somewhere). Once you have done this, the effects of bias current are cancelled, and only the "offset current" (the difference in bias current at the two op-amp input terminals) remains as an error. Measure I_{os}, the offset current, by looking at the residual dc level at the output; compare with specs: $I_{os} = 0.02\mu A$(typ), $0.2\mu A$(max).

8-2. Active Rectifier.

Construct the active rectifier in figure 8.4 (text figure 3.24). Note that the output of the circuit is not taken at the output of the op-amp. Try it with relatively slow sine

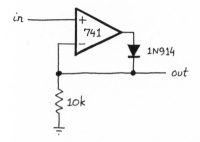

Figure 8.4. Active half-wave rectifier.

waves (100Hz, say).   Look closely at the output: What causes the "glitch"? Look at the op-amp output -- explain. What happens at higher input frequencies?

Now substitute a 355.   How is the performance improved?

8-3. Improved Active Rectifier.

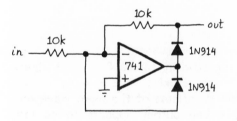

Figure 8.5. Better active half-wave rectifier.

Instead of the "brute force" solution (high slew-rate op amp) to the problem in the last exercise, try the clever circuit solution in figure 8.5.   Explain the improved performance.

8-4. Active Clamp.

Try the op-amp clamp circuit in figure 8.6.   (Note again that the circuit's output is not taken from the op-amp output; what significance does that have in terms of output impedance?). Drive it with sine waves at 1kHz, and observe the output. What happens at higher frequencies? Why?

Reverse the diode. What should happen?

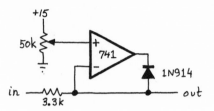

Figure 8.6. Active clamp.

8-5. Integrator.

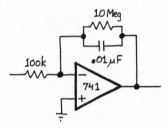

Figure 8.7. Integrator.

Construct the active integrator in figure 8.7 (text figure 3.44c). Try driving it with a 1kHz square wave. This circuit is sensitive to small dc offsets of the input waveform (its gain at dc is 100); if the output appears to go into saturation near the 15 volt supplies, you may have to adjust the function generator's **OFFSET** control. From the component values, predict the peak-to-peak triangle wave amplitude at the output that should result from a 2V(pp), 500Hz square wave input. Then try it.

What is the function of the 10Meg resistor? What would happen if you were to remove it? Try it. Now have some fun playing around with the function generator's dc offset -- the circuit will help you gain a real gut feeling for the meaning of an integral!

8-6. Differentiator.

Figure 8.8 is an active differentiator. Try driving it with a 1kHz triangle wave. Note: Differentiators are inherently unstable, because a true differentiator would have an overall 6dB/octave rising response; as explained in §3.33, this would

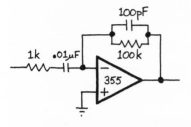

Figure 8.8. Differentiator.

violate the stability criterion for feedback amplifiers. To circumvent this problem, it is traditional to include a series resistor at the input, and a parallel capacitor across the feedback resistor, converting the differentiator to an integrator at high frequencies.

Reading: Chapter 3.20 to end, pp 122-147.
                  Chapter 4.03, pp 150-152.

Problems: Problems in text.
                  Additional Exercise 5.
                  Bad Circuits H,J,L.

9-1. Single-supply Op Amp.

The 358 dual op-amp (also available as a "quad" -- the 324)
can operate like any other op-amp, with V_+ = +15, V_- = -15;
however, it can also be operated with V_- = GND, since the
input operating common-mode range includes V_-, and the
output can swing all the way to V_-.

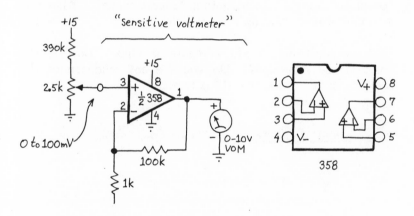

Figure 9.1.   Single-supply millivoltmeter.

   Try it out by constructing the "sensitive voltmeter"
circuit above. It has a full-scale sensitivity of 0.1V, and is
simply a non-inverting dc amplifier with a gain of 100.
Verify that it operates sensibly.

   Now try the same thing using a 741 or 355 (remember
that the pinout is different). Verify that the circuit will not
work with V_- = GND, but will work properly with the
conventional V_- = -15V.

9-2. Comparator.

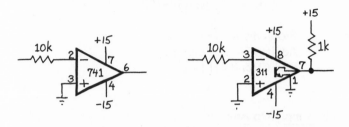

Figure 9.2. Simple comparators. The pin numbers shown in the figure are for miniDIP or TO-5 case; the pinouts for the awkward 14-pin DIP are different.

Try using a 741 as a **comparator**, as in figure 9.2. Drive it with a sine wave in the 1-10kHz range, and observe the (generally unsatisfactory) output 'square wave'. What op-amp limitation is causing this problem?

Now substitute a 311 comparator chip. How is the performance improved? Do you see any evidence of the oscillations for which the 311 is famous?

9-3. Schmitt Trigger.

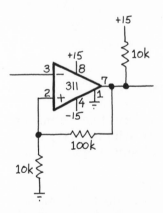

Figure 9.3. Schmitt trigger.

Connect a 311 in the Schmitt trigger configuration of figure 9.3, and predict the trigger thresholds.

a) Now drive it with a sine wave. Observe that triggering stops for sine waves smaller than some critical amplitude. Explain. Measure the hysteresis. Observe the rapid transitions at the output, independent of the input waveform or frequency. Look at both comparator input terminals; what happened to Golden Rule 1?

b) Reconnect the **"GROUND"** pin of the 311 (pin 1) to -15, and look at the output. What does the **"GROUND"** pin really do? The schematic diagram provided by the manufacturer (see, e.g., the National Semiconductor Linear Data Book) shows it as the emitter of the open-collector NPN output transistor, operated as a saturated switch.

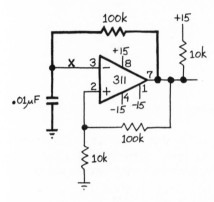

Figure 9.4. **RC** relaxation oscillator.

c) Now connect an **RC** network from output to input, as shown in figure 9.4. What should happen? Look at the output, then look at the inverting input (point **X**). (Again, what about Golden Rule 1?) Note that this circuit has no <u>input</u>. Can you predict the frequency of oscillation?

9-4. Negative Impedance Converter.

Connect up a negative impedance converter (figure 9.5, text figure 3.84). What should its input impedance be?

Now use it as the lower leg of a voltage "divider", as shown in the figure. What should the <u>gain</u> be?! Drive it with a 1kHz sine wave (1V pp, say), and see what comes out. Then try it with a 12k, or 11k, resistor instead of 15k. (This method of amplification has been used on telephone circuits, to amplify signals going in both directions on the same line.)

Note that interchanging the two op-amp inputs also results in a -10k synthesis. The difference is that the present circuit is stable only for driving impedances greater than 10k, while the NIC with the inputs interchanged is

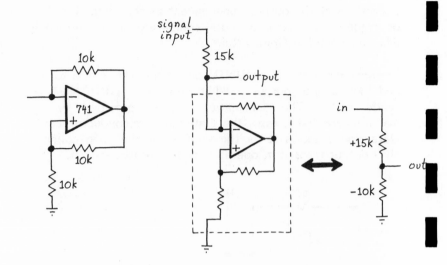

Figure 9.5. Negative impedance converter.

stable only for driving impedances less than 10k. You might enjoy wiring up the other variety, then driving it with a variable voltage source (of a few volts dc) while measuring the input current. What do you expect, given that the input to this NIC should behave exactly like a -10k resistor?

Reading: Chapter 4. For this lab the important part is
                4.11 - 4.16, pp 162-171. Read the first part
                of the chapter for general cultural
                enlightenment only.

Problems: Problems 4.4 and 4.5 in text.
                Additional Exercises 3,4.

10-1. IC Relaxation Oscillator.

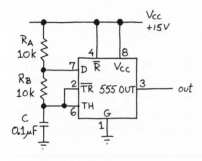

Figure 10.1. 555 relaxation oscillator.

Connect   a   555   in   its   classic   relaxation   oscillator
configuration   (Fig   10.1).   Look   at   the   output.   Is   the
frequency correctly predicted by

$$T = 0.693(R_A + 2R_B)C,$$

as derived in Exercise 4.5?   Now look at the waveform on
the capacitor.   What voltage levels does it go between?
Does this make sense?

   Now replace R_B by a short circuit. What do you expect
to see at the capacitor? At the output?

   Restore R_B = 10k, and replace R_A by a 470 ohm resistor.
What do you expect at the capacitor? At the output?

   Finally, try V_{CC} = +5V, to see to what extent the output
frequency depends on the supply voltage.

10-2. Sawtooth Wave Oscillator.

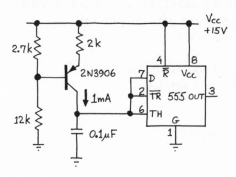

Figure 10.2. 555 sawtooth wave oscillator.

Generate a **sawtooth** wave by replacing R_A and R_B by a current source, as in figure 10.2 (and Additional Exercise 3). Look at the waveform on the capacitor (be sure to use a **10X** scope probe). What do you predict the frequency to be? Check it. What should the "output" waveform (pin 3) look like?

10-3. Triangle Wave Oscillator.

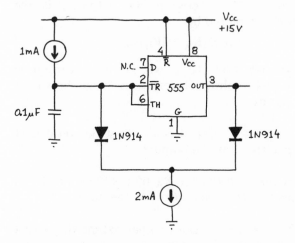

Figure 10.3. 555 triangle wave oscillator.

Use a 555 to make a **triangle** wave, as suggested in figure 10.3 (text figure 4.33). For the 1mA current source use the same circuit as in the preceding circuit; for the 2mA sink use a 2N3904 (NPN) with a 1k emitter resistor and a similar base voltage divider. Finally, look at the waveform on the 0.1μF capacitor, using a **10X** scope probe.

10-4. Voltage-Controlled Oscillator.

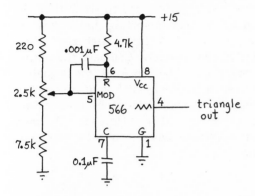

Figure 10.4.   566 VCO.

Try out the original VCO, the 566 (figure 10.4).   The specifications claim a minimum 10:1 frequency range, with the control voltages arranged as shown.   The output frequency is proportional to the voltage between the modulation input and the positive supply, with maximum frequency corresponding to $V_{mod} = 0.75 V_{CC}$.   The resistor on pin 6 and the capacitor on pin 7 set the overall frequency range.

10-5. Wien Bridge Oscillator.

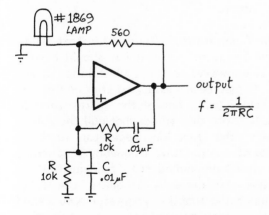

Figure 10.5.   Wien bridge oscillator.

Curiously enough, one of the most difficult waveforms to

achieve in an **RC** oscillator is the sine wave. The Wien bridge circuit (figure 10.5) accomplishes this by reducing the forward gain via an amplitude-dependent resistance (a type 1869 miniature lamp, rated 10V, 14mA) until the oscillation barely persists. Wire up the circuit, and look at the output waveform. It is relatively straightforward to show that the **RC** network from output to non-inverting input has an attenutation of precisely 1/3 at $f = 1/2\pi RC$, with zero phase shift (see if you can derive that result). Compare the output frequency with the prediction. Then look at the signal at the op-amp's inputs, to see if it checks with your prediction. Reducing **R** should reduce the amplitude of oscillation; try decreasing the value of **R** by 20%. Note the rubbery behavior of the output amplitude when you poke the non-inverting input with your finger.

10-6. Twin-T Notch Filter.

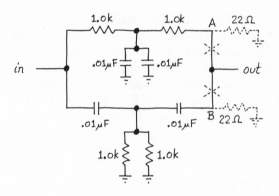

Figure 10.6. Twin-T notch.

Try wiring up a twin-T notch filter (figure 10.6, text figure 4.18). Calculate the frequency at which it should have a null. Try it out with sinewaves on the 100kHz range of the function generator. It may be necessary to trim one of the component values slightly (e.g., one of the 1k resistors) to obtain a complete null. Note that you cannot null the output signal entirely, since the function generator produces distortion at multiples of its operating frequency ("harmonic distortion"). After you have nulled the "fundamental" as completely as possible, by carefully trimming the filter components and retuning the function generator, you should be able to see the distortion products (i.e., the non-sinusoidal component) of the function generator.

In order to see how a simple **RC** network like this manages to produce infinite attenuation, break the connections and insert a pair of 22 ohm resistors as indicated by dashed lines in the figure. Then use both channels of the scope to look at the signals at points **A** and **B**. The small resistors let you look at the currents normally present when the filter is near null. As you approach that frequency, you should see the two currents become equal in amplitude and opposite in sign, resulting in a complete cancellation of output signal in the normal twin-T configuration.

Reading: Chapter 5.01 - 5.06, 5.10 - 5.12, 5.15 - 5.18,
                pp 172-83, 187-92, 199-204.

Problems: Problems in text.
                Additional Exercises.
                Bad Circuits.

11-1. Discrete 5V Regulator

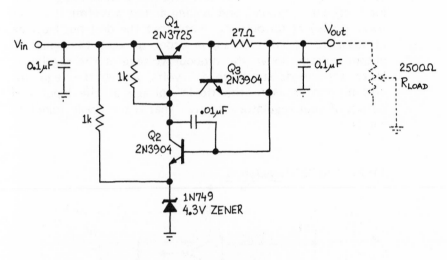

Figure 11.1. Discrete 5V series regulator. The
capacitors are needed to prevent oscillation.

a) Begin by building the +5 volt regulator in the figure
above (similar to text figure 2.73).   Watch how V_{out} varies
as V_{in} is varied from 0 to +20 volts (use the variable
regulated supply).   Connect the 2.5k potentiometer as a
variable load, to see how V_{out} is affected. What should the
maximum output current (I_{limit}) be?. Measure it.   Spend
some time studying the regulator circuit, to make sure you
understand the function of every component.

b) An important measure of voltage regulator performance
is its ability to reject ripple at the input.   To test this
'ripple rejection' property, connect a 6.3Vac filament
transformer in series with the variable dc supply, as in figure
11.2. The 33 ohm resistor and diode are included to prevent
damage in case V_{in} is accidentally shorted to ground; this
sort of connection, where a stiff voltage source can be
applied to the output of an instrument, can easily blow things

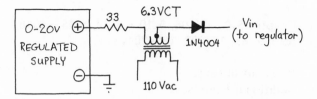

Figure 11.2. Ripple rejection test circuit. Note use
of "center-tap" connection to obtain 3.2Vac.

out. First, check out this test circuit by connecting a 2.2k
load resistor to ground, and looking at the waveform (dc plus
60Hz ripple) at its output. Note that the dc input must be
at least 12-14 volts or so, in order to keep the minimum
output voltage above the dropout voltage of the regulator.
Next, set the dc supply to +15 volts, attach the regulator
circuit, and measure the ripple voltage at the input and
output of your regulator circuit. What is the ripple rejection
ratio?

11-2. The 723 Regulator.

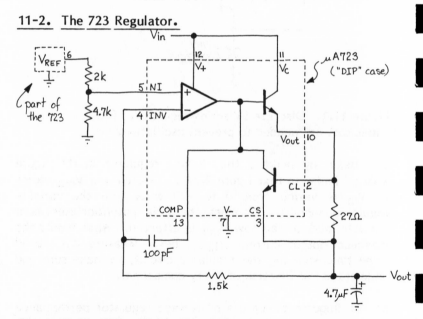

Figure 11.3. 723 positive 5V regulator.

This little chip does everything our previous circuit did, and
it does it better. Connect the regulator in figure 11.3
(similar to text figure 5.4). Connect the 0-20 volt supply to
V_{in} (unplug the 6.3Vac transformer, for now; it's OK to leave
it in the circuit). Measure V_{out}, and its variation with load
and V_{in} (called "line"). You should obtain superior
performance compared with the last circuit.

Plug in the 6.3Vac transformer, and measure the ripple rejection, taking care to provide sufficient dc input voltage. Convert the rejection ratio to dB, and compare with the 723 specifications in Table 5.8 (text p201). Calculate, for both this circuit and the previous one, the power dissipated by the regulator when V_{in} = 20V and the output is shorted to ground.

11-3. Three Terminal Fixed Regulator.

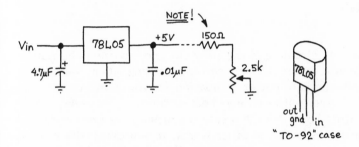

Figure 11.4. Three-terminal fixed 5V regulator.

Although the 7805 3-terminal positive regulator will not deliver the precision performance of the 723, it is a heck of a lot easier to use. Hook up the circuit in figure 11.4, which uses the lower current 78L05, and try it out. Reduce the dc input voltage until it drops out of regulation (turn off your ripple machine). What is the dropout voltage? Compare with the specifications in Table 5.7 (text p200).

Measure the ripple rejection, as before. Compare with specs (Table 5.7). Note the effect of dropout as the dc input voltage is reduced, with constant ripple amplitude. What limits does this set on the unregulated voltage applied to such a regulator?

11-4. Three Terminal Adjustable Regulator.

This variety of regulator is particularly convenient, being adjustable (you only need to keep one type in stock) and easy to use. Wire up the circuit in figure 11.5. Try **R** = 750 ohms; what should V_{out} then be? Measure it.

Replace **R** with a 2.5k pot, and check out the 317's performance as an adjustable regulator. What is the minimum output voltage (**R** = 0)?

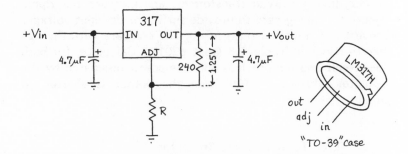

Figure 11.5. Three-terminal adjustable positive
regulator.

In this circuit and the previous one, shorting the output
will cause internally-set current limiting at a value that
exceeds the safe power dissipation for these low-power case
styles, used without any heat-sinking.   However, these
regulators boast "on-chip safe operating area protection and
thermal shutdown"; in other words, if you cause the chip to
get too hot, it should turn itself off rather than blowing
itself out.   You might wish to try this feature out, if you
dare!

11-5. Three Terminal Regulator as Current Source.

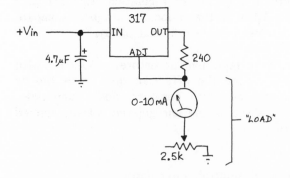

Figure 11.6. Simple current source made
with a 3-terminal regulator.

The 317 maintains 1.25V between its **OUT** and **ADJ** pins,
with very low current at the **ADJ** pin. Thus the "poor man's
current source" (Fig 11.6). Try it out. What should the
output current be?   Check its constancy as the load
resistance is varied.   How does the circuit work?   What
limits its performance at extremely low or high currents?

Reading: Chapter 6.01 - 6.11 (6.07,6.09 optional),
pp 223-242

Problems: Problems in Text.
Bad Circuits A,C,E,G,H,I.

12-1. FET Characteristics.

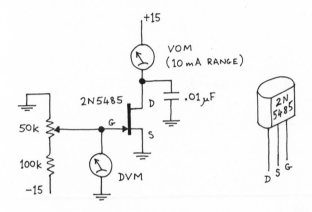

Figure 12.1. FET test circuit. Plot I_D vs V_{GS}.

Measure I_{DSS} and V_P ("pinch-off" gate voltage = V_T for a JFET) for two or three samples of 2N5485. Verify the relation between I_D and V_{GS} of figure 6.9 in the text. Notice the spread of values even in specimens from the same manufacturer's batch. Check that your values fall within the quoted maximum range:

$$4mA < I_{DSS} < 10mA$$
$$-4V < V_P < -0.5V$$

12-2. FET Current Source.

How good a current source is this? Investigate the lower end of the compliance range. What is V_{DS} when the constant current behavior starts to break down? This marks the boundary of the "linear region" and should occur when V_{DS} is near $-V_P$; try a FET with a different V_P to check this. Can you see how a FET could be used as a **2 terminal** current source (i.e. one that requires no external bias)?

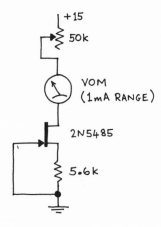

Figure 12.2. FET current source.

12-3. Source Follower.

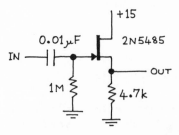

Figure 12.3. Source follower.

a)   Drive the source follower of figure 12.3 with a small sine wave of 1kHz. Look carefully at the input and output to see how much the gain differs from unity. Why? What is the maximum signal level you can get at the output without distortion? What is the limiting factor?

b)   Modify the circuit to include a current source load as shown in figure 12.4. You will find this to be a much better follower. Measure the DC offset; then interchange the FETs and measure it again. Explain. Measure the gain with a 1V, 1kHz signal (it had better be 1.0!). Attempt to measure the input impedance.

c)   Finally try the same circuit with a 2N3958 dual FET (figure 12.5). What is the offset of this follower? Is it within the specified upper limit of 25mV?

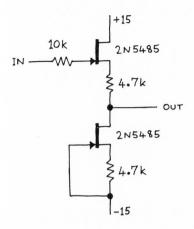

Figure 12.4. Source follower with current source load.

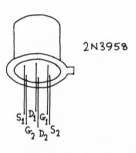

Figure 12.5. 2N3958 matched dual FET. The six lead package is most easily inserted into the breadboard as two rows of three (straddling the central median).

12-4. Voltage Controlled Gain Amplifier.

a)   Construct the circuit of figure 12.6.   Use a **small** (<50mV) input signal of 1kHz.   Measure the gain for a few values of the control voltage;   you should be able to vary this from +1 to +100.   How large a signal will it amplify without gross distortion (check this at several gain settings)? Find the corresponding V_{DS} values and explain the reason for the distortion and its approximate shape.

b)   Now try the improvement shown in figure 12.7 where the effects of V_{DS} on the channel resistance are nulled with a clever trick.   See §6.10 (p240) for explanation.   You should now see a dramatic improvement in performance.

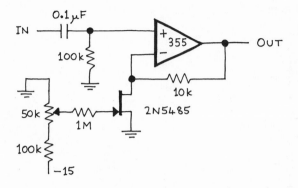

Figure 12.6. Voltage controlled gain amplifier.

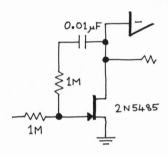

Figure 12.7. Circuit cure for distortion caused by
non-negligible V_{DS} effects. See figure 6.30 in text.

12-5. Automatic Gain Control (AGC).

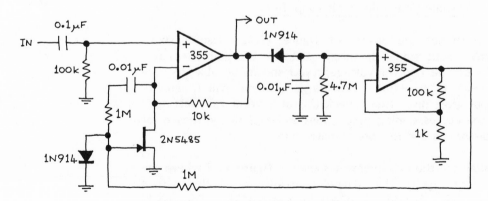

Figure 12.8. Automatic gain control circuit.

The voltage controlled amplifier you have just built is the basis of the automatic gain control (AGC) circuit of figure 12.8. The AGC has widespread use in radio and audio circuits for bringing a signal of unknown (or varying) level to some standard value. An example is the automatic level control in some tape recorders.

The AGC consists of a voltage controlled amplifier driving a peak detector and integrator with a decay time constant much longer than the period of the input. The integrated voltage, which is a diode drop below the level of the output (with negative polarity), is amplified by the second op-amp to feed back into the voltage control input of the first amplifier. Thus, the gain is adjusted until the output has a peak voltage of about one diode drop, which is a constant level.

Try out the circuit of figure 12.8. It should produce a 1V output for inputs in the range of 40mV to 1V without noticeable distortion. Look at the output of the second op-amp as you vary the input and observe its behavior as you wander out of the working range at either end. Explain.

Reading: Chapter 6.12 to end (6.17 optional),
                pp 242-261.

Problems: Problems in text.
                Bad Circuits B,D,F.

13-1. JFET Switch.

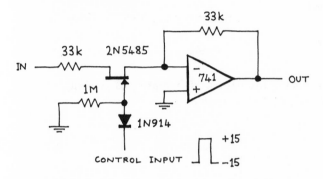

Figure 13.1. JFET Switch.

Try it with a 1V 1kHz signal, connecting the control input to
either +15 or -15. Why is the diode necessary? Would this
circuit work with the op-amp in the non-inverting
configuration?

13-2. CMOS Inverter.

Though classified as a logic circuit (we will look at this side
of it next time), the CMOS inverter has interesting properties
as a linear circuit too. It consists of a pair of
complementary MOSFETs connected as in figure 13.2.
Before you play with one, **read the handling precautions**
described on p249 of the text.

a) Now try one of the six inverters in a 4069 package with
the test circuit of figure 13.2(d). First tie all the unused
inputs (e.g. pins 9,11,13) to GND and connect V_{DD} to +15
and V_{SS} to GND. Watch the output voltage with a
voltmeter as you vary the pot; check that it agrees with
figure 6.51 of the text.

b) Now build the level shifting circuit of figure 13.3. Note
the unusual power connections to +15V and **GND**; this serves

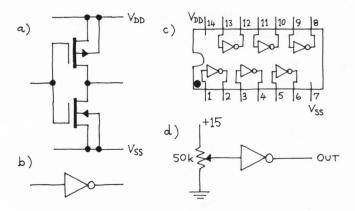

Figure 13.2. CMOS Inverter a) schematic, b) symbol, c) pinout (4069), d) test circuit.

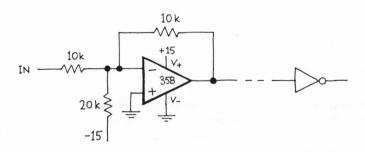

Figure 13.3. CMOS inverter driven by a 355 level shifting circuit.

to protect the CMOS circuits you will be using from negative input voltages. Give it an input from the signal generator (symmetric about zero) and increase the amplitude until the onset of clipping. Now look at the input and output of the CMOS inverter with the X vs Y mode of the scope so that you can reproduce the diagram of figure 6.51 (text). **Save the level shifter for the rest of the lab and always use it when you drive CMOS inputs from the function generator.**

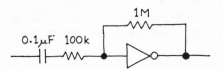

Figure 13.4. CMOS linear amplifier.

c) Wire up the inverting amplifier of figure 13.4. What is

the output quiescent point?   Can you predict that from figure 6.51?   What is the small signal gain?   What does this imply for the **open loop** gain of the inverter?   Hint:

$$G = A/(1+AB)$$

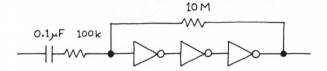

Figure 13.5. CMOS linear amplifier with three stages.

d)   Predict what will be the gain of the amplifier of figure 13.5.   Try it.

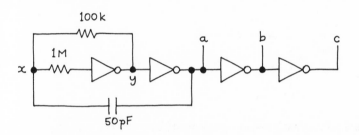

Figure 13.6. CMOS oscillator.

e)   Figure 13.6 shows a CMOS relaxation oscillator. Build it, look at the waveforms at 'x' and 'y', and compare the outputs 'a','b' and 'c'. Explain. Measure the frequency and make a note of it, and **save the circuit for later use.**

13-3. Transmission gate.

The 4066 contains four "analog transmission gates". Each is a fully buffered bidirectional switch which acts like a small resistance when the control input is at V_{DD} and an open circuit when the control is at V_{SS}. The only limitation is that the input and output must always lie between V_{SS} and V_{DD}, GND and +15V here. The gate is described more fully in §6.12 (p243); we will spend the rest of this lab looking at its uses.

a)   Set up the circuit of figure 13.7; connect V_{DD} to +15, V_{SS} to GND, and ground all unused logic inputs. Use the function generator combined with the level shifter above as

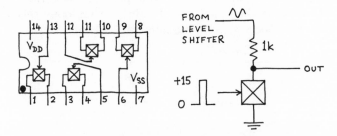

Figure 13.7. 4066 pinout and ON resistance test circuit. Note that the symbol is symmetric: the input and output are reversible.

input; use a 1kHz sine wave of several volts amplitude. Verify that the signal is switched on and off, and measure the ON resistance, treating the circuit as a voltage divider.

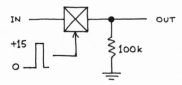

Figure 13.8. Transmission gate switch.

b) Now try the switch in the normal configuration of figure 13.8. The choice of a 100k load resistance is much better than the 1k used before (chosen so you could measure easily the ON resistance) and you should find this switch much cleaner. Attempt to measure the 'OFF' resistance by increasing the 100k resistor suitably. When you think you have an answer, try changing your input to "square wave" or varying the frequency. Explain.

13-4. Chopper Circuit.

Two transmission gates can be used to alternate signal sources at some high frequency (e.g. 30kHz that you have available from 13-2(e)) to give you the impression of two superimposed traces on your scope. This is in fact how the 'CHOP' mode of the scope works. For your two inputs, try looking at the function generator main output through the level shifter and at the 'sync out' of the generator, which is a 0-5V signal, hence within the input range. How should you trigger the scope to avoid seeing the chopping transients? How could you modify this circuit to offer 'select A' and

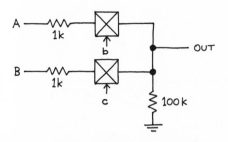

Figure 13.9. Chopper circuit.
The control inputs are driven from outputs 'b' and 'c'
of the 30kHz oscillator in figure 13.6.

'select B' functions?   What about 'ALT'?

13-5. Sample and Hold.

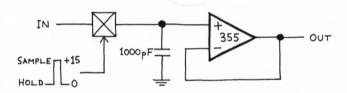

Figure 13.10.   Sample and hold.

A "sample and hold" is a switch circuit whose output follows
its input for one state of the control variable, and retains the
last held value of the input for the other state (see §6.14,
p246).   Construct the circuit of figure 13.10 and drive it,
through your level shifter, with a large amplitude signal of a
low enough frequency that you can follow it with the
voltmeter.   Connect the control input first to +15, then to
GND, and watch the sample-and-hold action (you may at the
same time see the 'memory' effect of an open circuit CMOS
input).   Because the "hold" function relies on retaining a
charge stored in the capacitor, this circuit is prone to
"droop".   Measure the droop rate and try to estimate the
asymptotic voltage; from this you may be able to determine
whether the gate or the op-amp is causing the droop.   The
choice of capacitor is a compromise between speed (output
slew rate) and droop rate; here, the capacitor is chosen for
maximum **speed**, so the droop rate could be made much
lower, if desired.

13-6. Commutating Filter.

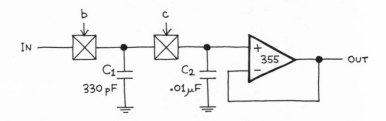

Figure 13.11. Commutating filter. Inputs 'b' and 'c'
are connected to the oscillator of figure 13.6.

Two consecutive sample and holds connected as shown in
figure 13.11 act as a low pass filter. Refer to §9.40 (p448)
for a derivation of the formula for the -3dB point and a
detailed explanation. In simple terms, during each cycle of
the sampling clock (here running at 30kHz), C_1 dumps a
charge proportional to the difference between the input and
output voltages onto C_2. Because C_1 is much smaller than
C_2, it takes many such cycles for the voltage on C_2 to
approach that of the input. If the input itself is changing in
time at anywhere near the rate of the sampling clock, the
output is diminished in amplitude, and the net result is a 6dB
per octave rolloff.

Verify the frequency response and compare the measured
value of the -3dB point with its calculated value.

Reading: Chapter 8.01 - 8.15, pp 316-341.

Problems: Problems in text.
Additional Exercises 11,12,14,15.

Now we start the second half of the laboratory course with the first of the labs on logic circuits. First a word of warning.

CAUTION. In all work with logic circuits:

1) Never apply > 5V or < 0V to the LED indicator lights on the breadboard.

2) Never run TTL circuits from any supply other than +5V and GND.

Violating these rules results in certain destruction of various IC's including the ones inside the breadboard box itself.

14-1. Logic Probe and Level Indicators.

a) Investigate the switch outputs and LED indicators on your breadboard. Connect a plug lead to the terminal of one of the red LED indicator lights on the breadboard. Touch it to +5V, then to ground. Now connect it to one of the breadboard switches with a 'LEVEL'[†] output and toggle the switch. Look at the LEVEL output with a voltmeter to see the TTL voltage levels.

† Your breadboard may have switches that produce simply +5V or GND instead of TTL levels; if so, you will discover it with this test.

b) Now do the same thing with a logic probe. Attach it to a breadboard BNC connector and apply +5V to this. Some logic probes are designed to work with CMOS, also, and will tolerate a 15V supply; make sure you find out beforehand because **you will certainly destroy a 5V logic probe with 15V.** Notice that when it is connected to an open circuit it is at half brightness[††] so that you can distinguish it from both of the logic states.

†† Your logic probe may have some alternative way of indicating a floating input.

c) Connect both the level indicator and logic probe to the push button output marked 'PULSE' (if you have one) and push the button. What is going on? You can verify that a pulse is there if you like with the oscilloscope, but you will find that it is very short, too short in fact to see with the LED. This demonstrates the pulse stretching capability of the probe. A good probe can pick up pulses of either polarity as short as 20ns, so is very useful for tracing glitches.

d)   Finally connect the LED and logic probe to the TTL breadboard clock marked '1KHz'. Here you see the probe giving a flashing display (as if it were stretching both the highs and lows), while the LED flashes at 1KHz and looks to you as continuously dim.

14-2. Diode Gates.

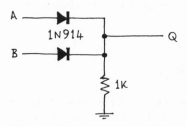

Figure 14.1. Diode OR gate. Inputs are A and B, output is Q.

a)   Diode OR gate. Feed the LEVEL outputs into A and B and look at Q with a voltmeter and with an LED.

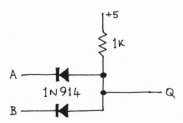

Figure 14.2. Diode AND gate.

b)   Diode AND gate. For this circuit, check all four entries in the truth table. What are the output voltage levels (both low and high) of these circuits, when you drive them with inputs of +5.0V and 0.0V (GND)? What is the 'fanout' of these diode gates?

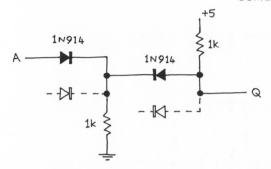

Figure 14.3. Cascaded OR and AND configurations.

c)   Do the same for the series combination of figure 14.3.
Why couldn't you build a large logic circuit entirely from
these 'M^{2}L' (Mickey Mouse Logic) gates?

14-3. Discrete TTL NAND Gate.

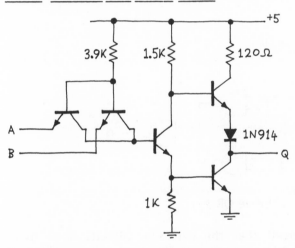

Figure 14.4. Discrete TTL NAND gate.
All transistors are 2N3904. The input protection
diodes are omitted for simplicity.

Construct the circuit of figure 14.4. Drive the input from
the breadboard LEVEL outputs, and verify that the output is a
true NAND function of them. Measure the actual high and
low output voltages. Explain how this circuit works.

Convert your circuit to that of figure 14.5. First tie the
E input (ENABLE) high and verify that it behaves as before.
Then tie E low and look at the output with your logic probe.
While it is in this disabled state attach another TTL
output (such as a breadboard switch with a LEVEL output) to
Q and check that the new level overrides.

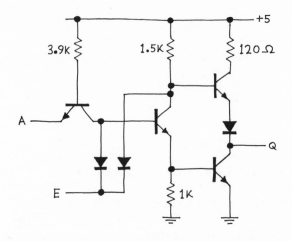

Figure 14.5. Discrete TTL inverter with tristate output. All transistors are 2N3904, all diodes 1N914.

14-4. Exclusive OR.

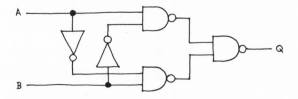

Figure 14.6. Exclusive OR #1.

Convince yourself that the exclusive OR circuit of figure 8.23 in the text is the same as figure 14.6 here, where all the gates have been substituted with their NAND equivalents. Construct it with TTL integrated circuits, 74LS00 (quad NAND) and 74LS04 (hex inverter), and try it out. Use the pinout appendix at the back of the book for the connection diagrams. Get used to the conventional choice of pins for V_{CC} (+5V always for TTL) and GND in opposite corners of the package. Also, if you get into the habit of pointing all the chips the same way on the breadboard, the power supplies become very easy to wire up.

Now check the circuit in figure 14.7 is equivalent to that of 8.24 in the text, and construct that one too. You might consider this as an experimental verification of de Morgan's theorem. Can you think of a way to realize exclusive OR with just a single package of NAND gates?

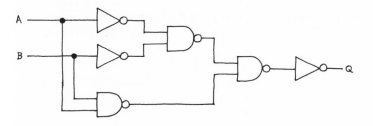

Figure 14.7. Exclusive OR #2.

14-5. Multiplexer.

Get a 74LS151 "8 input multiplexer" circuit and investigate its operation. Connect an LED indicator to its output Q (use the pinout appendix), set up an address on inputs A,B,C and ground the STROBE ($\overline{\text{ENABLE}}$) input, then momentarily bring each one of the data input lines D_0 to D_7 to ground (recall that an open TTL input is seen as logic high) in turn until you find out which one of them is 'live'. Do this for several addresses until you understand the function.

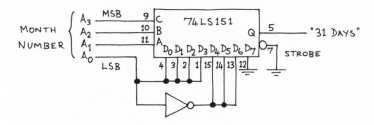

Figure 14.8. 31-day machine. Pin connections other than V_{CC} and GND are labelled.

Now construct the 31-day machine of figure 14.8, which is figure 8.38 of the text. The circuit lights up the LED when the month whose number is applied to the input lines has 31 days in it. Check it out by connecting breadboard LEVEL outputs to the least significant bits of the binary month number and wires to the rest, comparing with your knuckles!

14-6. Adder and Magnitude Comparator.

The 74LS83 "4 bit full adder" performs the parallel binary addition of one 4 bit number (A_1 to A_4) to another (B_1 to B_4), including a "carry in" bit (C_0), to produce a 4 bit sum

(S_1 to S_4) plus a "carry out" (C_4). Use it to generate excess-3 outputs from binary inputs for the integers 0-9 (see §8.03, p321). What should you do with C_0? What should appear at C_4?

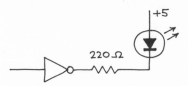

Figure 14.9. A way to drive a light emitting diode from TTL levels.

Now figure out how to add a 74LS85 "4 bit magnitude comparator" to flag illegal numbers, i.e. input > 9. For the cascading inputs, connect 'A<B' and 'A>B' to GND, 'A=B' to V_{CC}. For the output, use a fifth LED connected as shown in figure 14.9.

14-7. CMOS Gates.

You met CMOS circuits last time, but you should reread the handling precautions in the text p249 for safety. Also, check that your logic probe is suitable for use with CMOS (V_{DD} = 15V); **if not, put it away to avoid the temptation to use it!**

First, try out a 4011 quad 2 input NAND. See the appendix for the pinout. Use V_{DD} = 5V and V_{SS} = GND so that you can still use the LEVEL outputs and the LED indicators. Ground one input of each gate **that you do not use** as you did last lab, a precaution that is not necessary with TTL, and apply power. Look at the outputs with a voltmeter and observe the clean saturation at V_{DD} and GND. Do the same, driving with the 1KHz clock and looking with the oscilloscope.

Finally, try operating the gates from V_{DD} = 15V. To do this you will have to provide inputs from +15 or GND by direct connection, since the breadboard LEVEL outputs are 5V TTL levels. Look at the output with a voltmeter and note the clean saturation again.

LAB 15. SEQUENTIAL LOGIC I

Reading: Chapter 8.16 - 8.25, pp 341-359.

Problems: Problems in text.
              Additional Exercises 1-8.
              Bad Circuits.

15-1. JK Flip Flop.

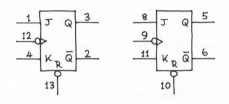

Figure 15.1. 74LS107 dual JK flip flop.
GND = Pin 7. V_{CC} = Pin 14.

a) Begin by examining the function of a 74LS107 dual JK flip flop. Clock it with positive pulses from the breadboard switches. Try all four levels of J and K, with $\overline{RESET}$ connected to V_{CC}, and write out a truth table. Compare with the advertized version on p345 in the text. Then try the $\overline{RESET}$ input.

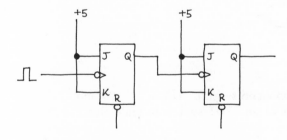

Figure 15.2. Cascaded JK flip flops.

b) Now cascade the two flip flops in the package, connecting Q_1 to clock 2; Connect J and K for toggling. Display the states of the two Q outputs on LED indicators. Satisfy yourself that it is a divide-by-4 circuit and that it counts the binary sequence 0,1,2,3,0...

15-2. Debouncing.

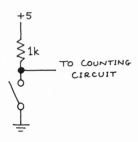

Figure 15.3. Undebounced switch. Get an SPDT
(single pole double throw) switch from stock for this;
do not try to use one on the breadboard.

a) Try clocking the cascaded flip flop circuit directly from a
switch as in figure 15.3. What does the 1K resistor do? Why
doesn't the counter function correctly when driven this way?
See the text p342 for guidance.

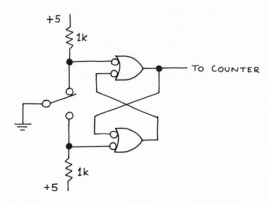

Figure 15.4. Debounced switch. Use a 74LS00 and
all three terminals of an SPDT switch to build this.

b) Fix the problem by adding a debouncing circuit. Note
the use of assertion level logic in figure 15.4; the gates are
NANDs. The counter should now step one count per cycle
of the switch. The same kind of circuit is frequently used
inside a breadboard for the push buttons; if yours does not
have this feature, you will have to build figure 15.4 every
time you need a debounced switch, and you should keep your
assembled circuit for the rest of this lab.

15-3. D-type Flip Flop.

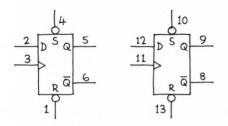

Figure 15.5. 74LS74 dual D-type flip flop.
GND = Pin 7. V_{CC} = Pin 14.

a) Remove the 74LS107 and try a 74LS74 dual D-type flip
flop. Tie both $\overline{SET}$ and $\overline{RESET}$ to V_{CC}, then check that the
data at the D input is clocked through to the Q output on
the rising edge of the clock. Now connect $\overline{Q}$ to D to make a
toggling flip flop. Make sure that this works.

Finally check to see whether SET and RESET outrank the
other inputs by asserting (connecting to GND) one of them,
then trying to toggle the flip flop. What happens if you
assert both SET and RESET simultaneously (be sure to look at
both Q and $\overline{Q}$ before you answer)?

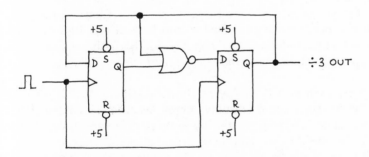

Figure 15.6. Synchronous divide-by-three with a
74LS74.

b) Wire up the synchronous divide-by-three circuit of figure
15.6 (figure 8.52 in text). The NOR can be either one
section of a 74LS02 (quad 2 input NOR) or else a 74LS00
(NAND) cleverly contrived to look like one. Set the counter
to the state that is never reached in the divide-by-three
sequence by use of the $\overline{SET}$ and $\overline{RESET}$ inputs and verify that
it does not hang in that state when toggled. Write down the
full state diagram. When you have checked the function
with your push button and lights, try it with the 1KHz clock

and oscilloscope. Draw the signals seen at Q_1, Q_2 and the input on the same sheet and make sure you see how these relate to the count sequence.

15-4. MSI Counters.

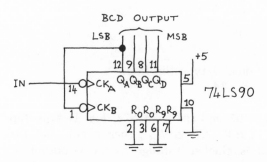

Figure 15.7. Decade counter, 74LS90. **Important** this chip has unusual power connections: V_{CC} = Pin 5. GND = Pin 10.

a) Connect a 74LS90 decade counter for BCD counting as in figure 15.7. Note that the chip consists of two sections, a divide-by-2 clocked by 'CK_A' with output 'Q_A', and a divide-by-5 (3 flip flops in a synchronous counter) clocked by 'CK_B' with outputs 'Q_B' to 'Q_D'. To make a divide-by-10 BCD counter you must connect them as shown. Clock the input with the breadboard push button and look at the four output bits with LED indicators. Then try out both the resets 'R_0' and 'R_9' and figure out what they stand for.

b) Now take an HP display as shown in figure 15.8. These are very convenient display chips as they combine the functions of latch (activated by $\overline{EN}$), plus hexadecimal (0-9, A-F) decoder/driver and LED array display. Connect the BCD outputs of the counter to the inputs A to D and count through the states. Finally try substituting a 74LS93 divide-by-16 for the 74LS90 (pin compatible, but without an 'R_9') and repeat.

c) Now look at the outputs Q_A to Q_D of the 74LS93 in turn with the scope while you apply the 1KHz clock to the counter input to verify that each is half the frequency of the previous one. Then do the same with the 74LS90 and justify what you see. How could you connect a 74LS90 to give you a **symmetrical** square wave output at one tenth of the clocking frequency?

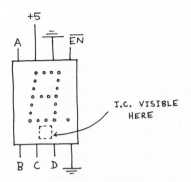

Figure 15.8. Hewlett Packard 5082-7340 ("HP")
display. These displays should be mounted in sockets
to avoid bending the fragile leads: **leave them that
way.**

15-5. Programmable Divide-by-N Counter.

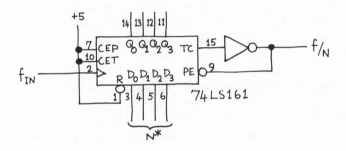

Figure 15.9. Divide-by-N using a 74LS161
synchronous binary counter.

The 74LS160-3 series of counters are 4 bit **synchronous**
binary (divide-by-16) and decade (divide-by-10) counters
with either synchronous or master (asynchronous) resets. All
of them have the feature of direct parallel synchronous
loading (i.e. loads on the rising edge of the clock) from
inputs 'D_0' to 'D_3' enabled by 'PE', "parallel enable". We
can exploit this feature to construct a divide-by-N
programmable counter with the circuit shown in figure 15.9.
N*, the 2's complement of N, is applied to the parallel inputs
which are then enabled when the output 'TC' goes high
("terminal count": 15 for a binary counter, 9 for a decimal
one); the counter is then loaded with N* and the sequence
starts again. Try out this circuit with various values of N.

15-6. Stopwatch.

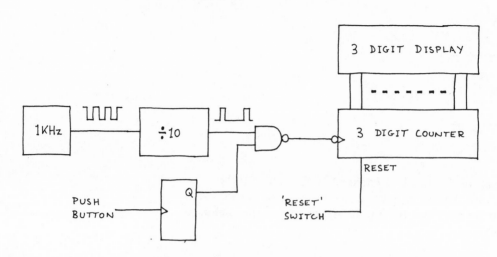

Figure 15.10.   Stopwatch block diagram.

The last exercise today is to build a stopwatch according to the suggestion of figure 15.10.   Use 100Hz (1KHz breadboard clock divided by 10), gated into a 3 digit counter with the output bits connected to displays.   Control the gate into the counter with the output of a toggling flip flop driven from the breadboard (debounced) push button.   Use a second switch to reset the counter.   Remember to figure out what to do with the various inputs: RESETs, CLOCKs, ENABLEs, etc.   This should all take about an hour, so don't start unless you have time to finish.

Reading: Chapter 8.26 to end, pp 359-379.

Problems: Additional Exercises 9,10,13.

In this lab we will explore tristate devices, and use them in a unique "monobus" circuit (**unique** because you would have to be crazy to build such a circuit in real life!)

16-1. TTL 3-State buffer.

We saw how the addition of a tristate ENABLE input was a straightforward modification to a TTL gate in lab 14.   Now look in the appendix at the pinout of the 74LS125 quad tristate buffer.   It has four independent non-inverting buffers each with a separate $\overline{\text{ENABLE}}$ input.   To assert the output, the $\overline{\text{ENABLE}}$ must be low; to switch it to the open circuit ("high Z") state, the $\overline{\text{ENABLE}}$ must be high.

        To try it out, first enable one section and verify that the output follows the input (use LED indicators and switches). Then disable it and check that the output can be brought high or low by connecting it to +5V or GND through a 2.2k resistor.

16-2. "Monobus" Example of a Data Bus.

For the rest of this lab you are going to construct a data 'bus' using tristate buffers. A bus is a set of shared lines that may be driven by one of several data sources, with some agreed upon protocol to prevent "bus contention".   Consult the text p330 (figure 8.19) for an introduction.   Chapters 10 and 11 are full of them, and we will meet them in later labs too.

        A simple, but non-trivial example of the use of a bus is shown in figure 16.1.   The 74LS90 at the bottom is driven by a push button, and is considered to be our source of data. To get the BCD count over to the display on the right hand side we use a 1-bit bus, putting each bit in turn onto the bus via the 74LS125 buffers in the order Q_D, Q_C, Q_B, Q_A.   This task is done at a rate of 1kHz by using a 2 bit counter (74LS93) and 1-of-8 decoder (74LS138), enabling each buffer in turn.

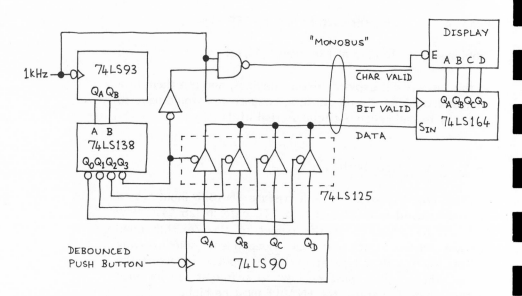

Figure 16.1. "Monobus" functional diagram. This time, it is left as an exercise for the reader to figure out the pinouts and what to do with the remainder of the inputs to each chip.

At the other end, a shift register (74LS164) picks off the bits, using the rising edge of BIT VALID as a clock. We only use the first 4 bits of the 8-bit chip. The display latches an updated BCD character using $\overline{\text{CHAR}}$ $\overline{\text{VALID}}$ once every four bits. Does it matter that the shift register is not reset after the display is latched?

Begin by getting out some paper and drawing a timing diagram, putting the 1kHz clock on the top line. Indicate the states of the 74LS93, the 74LS138 outputs and the three bus signals using a format something like figure 8.55B on p350 of the text. When you are sure there is no problem with bus contention or logic races, you are ready to build the circuit.

Now figure out what to do with all the inputs that are not specifically assigned in figure 16.1; these have been purposely left out to avoid making the exercise read like a cookbook. Consult the pinout appendix. In particular:

74LS93 CK_B clock input?
          R_0 inputs?

74LS138 C,D,ENABLE and $\overline{\text{ENABLE}}$ inputs?

74LS90 CK_B clock input?
         R_0 and R_9 inputs?

74LS164 RESET input?
         Second serial input?

   Try out your circuit. The chances are reasonably good
that you won't get it right first time. If so, devise a rational
troubleshooting procedure to find out what is wrong. Two
approaches are popular:

1)   Trace from the inputs (i.e. the clock and the push
button) chip by chip through the circuit toward the outputs
(i.e. display).

2)   Starting at the output, break the circuit between two
chips and insert logic levels to make sure the output (display,
here) responds as it should. Normally, do this for individual
lines, one at a time and work your way back toward the
input.

Reading: Chapter 9.01 - 9.08, pp 380-398,
                  9.18 - 9.19, pp 408-413,
                  9.22 (dual slope only), pp 415-418.

Problems: Exercise 9.1.
                  Additional Exercises 1,2.
                  Bad Circuits.

Today we are going to examine digital-analog techniques.
We will look at a 1408 8-bit current-switching DAC and build
a tracking ADC by using it in a feedback configuration.
**Wire your circuits neatly; each part builds on the one
before it.**

17-1. Digital-Analog Converter.

a) Begin by hooking up a 74LS191 4 bit synchronous binary
up/down counter. See the appendix for the pinout. Tie the
$\overline{\text{ENABLE}}$ input low and parallel $\overline{\text{LOAD}}$ high. Connect the
outputs to the indicator LEDs and clock the chip from the
debounced push button (use the 'normally high' output).
Check the count sequence in both the "up" ($\overline{\text{U}}/\text{D}$ = low) and
"down" ($\overline{\text{U}}/\text{D}$ = high) modes.

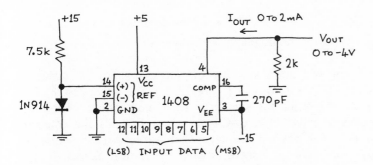

Figure 17.1. 1408 8-bit monolithic current switching
digital-analog converter. The current output is
converted to a voltage by a 2k resistor.
Note the polarity of the output.

b) Now use the output of the counter to drive a 1408 DAC
as in figure 17.1. Connect the counter's four output bits to
the **most** significant input bits (i.e. pins 8,7,6,5) of the DAC.
Which is the correct order? Clock the 74LS191 with the

1kHz breadboard signal. What waveform should you see at the DAC output? Look with the scope. Then reverse the direction of the counter to get the other "staircase" wave.

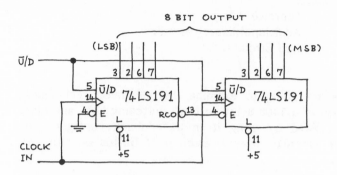

Figure 17.2. Fully synchronous 8 bit up/down counter using 74LS191s.

c) Add a second 74LS191 in the manner of figure 17.2. RCO ("ripple carry output") is used to generate the carry: pin 13 is low during the last count of the sequence (count 15 in count up mode, count 0 in count down); this is tied to the count $\overline{\text{ENABLE}}$ of the next stage which then steps synchronously at the next clock pulse. Tie all the bits to the 1408 (order?) and look at the new "staircase" wave, which is now more like a ramp.

17-2. Tracking Analog-Digital Converter.

a) If we add a comparator circuit we can now make a tracking ADC as shown in figure 17.3. See §9.22 (p416) for details. The 2.5k pot simulates an analog input. The 311 compares this with the DAC's output and clocks the counter up or down as needed to make that agree with the analog input. The counter's state is then the digital output of the DAC.

Display the four most significant bits on an HP decoded LED display as shown. Run the 2.5k pot slowly up and down through its range. What happens at the ends? Is this reasonable?

Now look at the voltage output of the 1408, which should always be dancing between two adjacent levels of the staircase. Then twiddle the pot rapidly and note the larger jumps, which only occur while you are actually turning the pot, not while the DAC is ramping towards its new state.

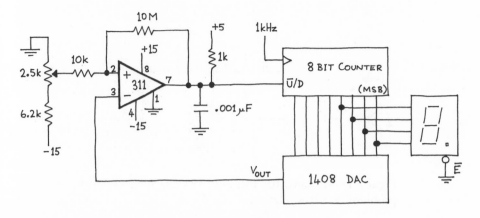

Figure 17.3. Tracking ADC using 8-bit counter and DAC of 17-1. The addition of hysteresis and capacitor (which slows down the slew rate) to the 311 comparator are to reduce its tendency to oscillate.

These jumps are caused by an error in the sequence of the 74LS191s that happens when the $\overline{U}/D$ input level is changed while the clock is low. The specification of the 74LS191 says that the $\overline{U}/D$ and $\overline{ENABLE}$ inputs must be stable during the entire time that the clock is low, and this rule can be broken if the comparator output is made to change **between** clocks by an analog input changing in time. Convince yourself that this is true by drawing a timing diagram for the clock and the comparator output.

b) To see the dancing behavior of the ADC in its steady state more easily, disconnect the four least significant bits driving the 1408. Now there will be only 16 levels on the scope, and the DAC output will always be oscillating between two of them.

c) A cure for the count sequence error problem we met in 17-2(a) is shown in figure 17.4. The comparator's output is sampled with an edge triggered flip flop (74LS74) at a time when the clock is guaranteed to be high. The monostable (74121) is used to delay the (rising) clocking edge by 70 microseconds to allow the DAC output and comparator to settle; it would be more elegant to omit this, but doing so would cause the ADC to "hunt" by 3 steps rather than just one. Explain this last statement.

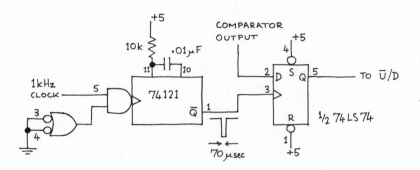

Figure 17.4. Cure for the 74LS191 clocking violation problem dicussed above.

Try the cure. If things don't work at first, check that the monostable is working. The ADC output should now be free from glitches.

Reading: Chapter 9.28 - 9.39, pp 428-446. **Don't worry about the mathematical details**; these are rather tricky.

Problems: Exercise 9.4.
                 Additional Exercise 4.

18-1. Psuedo Random Bit Sequence Generator.

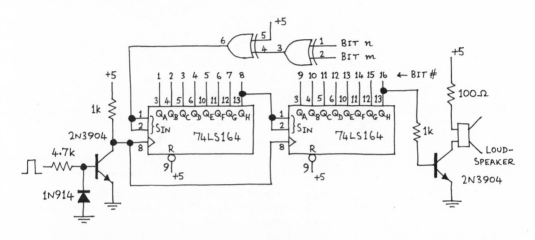

Figure 18.1. Pseudo ramdom sequence generator.
The 74LS86 and 74LS164 have the standard 'corner'
power supply connections.

This circuit is the standard PRBS generator of §9.34 with a maximum of 16 stages of shift register available. The second exclusive OR gate inverts the feedback, so that it can't get stuck in the state of all zeros which would be the unique stable state (the state of all ones is, instead); in this way, RESET can be used to start up the generator if necessary. Consider the evolution of the state of all ones to verify this statement (see p443).

    Use the function generator at the input to drive the clock with a large square wave of 100kHz. Begin by setting the taps to m = 15, n = 14 (a maximal length PRBS) and listen to the "noise" with the loudspeaker; it should sound "white", i.e. featureless or without any apparent frequency. The only periodicity present is the repeat length of the sequence. Calculate that repeat time. Can you hear it?

Next try m = 15, n = 16 and then m = 15, n = 13. Neither of these gives a maximal length sequence. Try other combinations too.

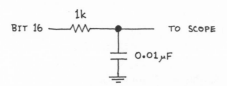

Figure 18.2. Low pass filter for PRBS generator.

Connect up the filter of figure 18.2 to one of the outputs. Why does it not matter which one? Run the clock at 1MHz now and look at the filtered waveform on the oscilloscope. Explain the general shape of what you see in terms of the clock frequency and the f_{3dB} of the filter. Trigger the trace carefully on the highest (or lowest) peak in the sequence; this is possible even for the maximal length sequence. Look with the other trace at the unfiltered output. Spot the longest sequence of consecutive 1's, also consecutive 0's. Explain these lengths in terms of the length of the shift register itself. What feature on the filtered output do these correspond to?

Now turn off the horizontal sweep of the scope, or run it very slowly. The filtered trace is now reduced to a vertical line of variable brightness. The brightness at each vertical position is proportional to the amount of time the output is spending at that particlar voltage. Describe in rough terms the distribution you see; what does it approximate?

Keep the transistor circuit for the next exercise.

18-2. Phase Locked Loop Frequency Multiplier.

Construct the circuit of figure 18.3. Set the function generator which drives the input to 60Hz exactly using the lab frequency counter in the period timing mode. What frequency should now be present ot the VCO output (pin 4 of the 4046)? Then measure it with the counter. Look at some of the intermediate outputs of the 4020 also.

Now look at the output of the phase detector (pin 13 of the 4046). It is a type II detector as described in §9.29 (p429). You'll notice a string of short (but not zero length) positive going pulses. The theory predicts that these should

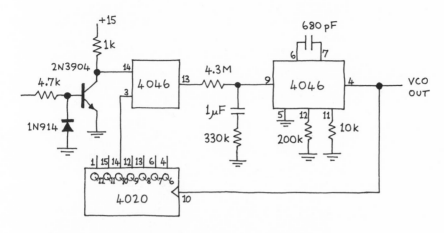

Figure 18.3. Phase locked loop frequency multiplier circuit based on the CMOS 4046 PLL. This is based on figure 9.58 of the text. Pin 16 = V_{DD} = 15V and pin 8 = V_{SS} = GND on both chips. The phase detector and VCO of the 4046 are drawn as separate boxes.

vanish in the steady state, however the 10 megohm load of the scope probe you are using is discharging the filter capacitor enough to make them visible. Interpose a 355 follower between the capacitor and the probe and they will go away. **Slowly** adjust the input frequency as you watch the detector output. Do things make sense now?

The time constant of the filter is now chosen correctly for a 60Hz input frequency. Investigate the effect of varying some of the values: in particular, replace the 330k resistor by 33k and watch the PLL hunt up and down before it settles on the right frequency. This is an impressive sight!

Lastly try out the type I detector that the 4046 also offers. Its output is pin 1, and the inputs are the same as for the type II detector, so simply move the wire from pin 13 to pin 1. You should be able to see the fluctuation of the VCO frequency over the period of the input, which you can exaggerate by reducing the size of the 1uF loop filter capacitor. If you make sudden changes in the input frequency you should also be able to lock onto its harmonics.

Reading: Chapter 10.1 - 10.11, pp 453-468.

Problems: Additional Exercises 1-6.

There will be a total of five labs in this series, each one
building on the last one. The computer is assembled in the
first two sessions and the last three are self-contained
experiments based upon it (which can therefore be taken in
any order). The subject of each computer lab is as follows:

Lab 19	Data and address buses. Memory.
Lab 20	Central processor. Testing and characteristics.
Lab 21	Timing methods and simple input/output. Decimal arithmetic.
Lab 22	Frequency meter.
Lab 23	Analog conversion methods. Simple Graphics.

Now look in figure 19.1 at the block diagram of the
computer we will build. We are using the popular Z80
microprocessor chip because it has static registers and so
allows us to use a slow clock to observe the execution of
individual instructions.   We will stick with the 8085
instruction set you meet in chapter 11 for convenience; this
is a subset of the entire Z80 instruction set and is completely
compatible.   If you ever refer to Z80 literature you will
have to get used to the alternative mnemonics (the op-codes
are of course the same!); if you do, you may like to take
advantage of the extended instruction set, but this will not
be necessary to complete these labs.   Before you begin, a
few general suggestions:

1)   Be very neat with the wiring. Keep the connections as
short as you can and color code them; you'll realize later
how helpful this is for finding mistakes.   It is particularly
helpful to use a different color for each bit of a bus as
accidental permutations of bits are a frequent mistake. You
might label them in accordance with the resistor color code,
black for bit zero etc.

2)   Pack the chips close together.   If you are frugal,
everything will fit onto three standard breadboard strips, two
for the computer itself and one more for the various
peripherals.

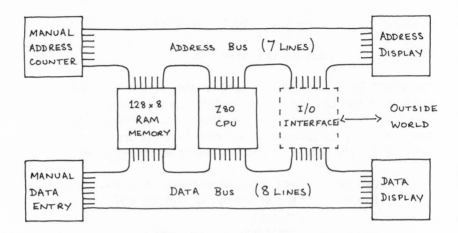

Figure 19.1. Computer block diagram.

3)   Handling of MOS chips. MOS circuits can be destroyed by static charges on the input pins. **Always** insert a MOS chip **after** connecting up its leads on the breadboard. **Never** touch the pins with your fingers; handle the package by its ends only, and store it always in conductive foam. This precaution applies only to the CPU and RAM chips in these labs.

19-1. Address Counter and Display.

Connected as shown in figure 19.2, the tristate buffers are redundant and the address bus is always asserted by the counter, so the display shows the state of the counter. Verify that you can count up and down with the push button and that spurious counts are not generated when you change the direction (if you have this problem, the polarity of the pulse from the push button is probably wrong).

Raise the $\overline{\text{ENABLE}}$ of the buffer to high. Explain the display. Now look at individual bus lines with the logic probe as you enable and disable the tristates. The logic probe is designed to show the floating state as half brightness. Leave the $\overline{\text{ENABLE}}$ grounded for the present; it will eventually be controlled by the computer, as will the $\overline{\text{ENABLE}}$ for the display (also grounded).

19-2. Data Bus and Display.

Construct the data bus structure shown in figure 19.3. Order the bits D_7 (MSB) D_6 ... D_0 (LSB) from left to right

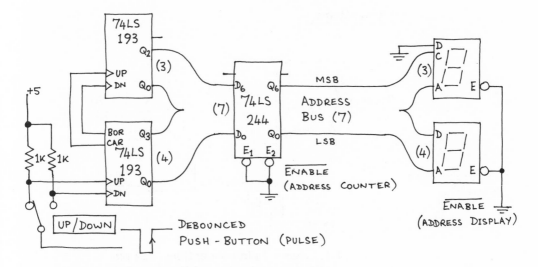

Figure 19.2. Address counter and display. Not all
the pins of the 74LS193 up/down counters are
assigned; what should you do with the unused inputs?
Consult the pinout appendix.

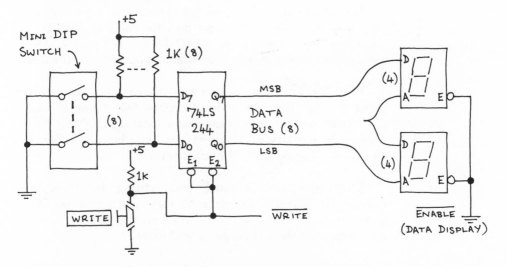

Figure 19.3. Data bus and display. The 'pinout' of
the mini-DIP switch is as indicated in the diagram.

as the breadboard faces you so that the bit pattern
represents a binary number; order the two displays similarly.
Check that when WRITE is low the display shows the bit
pattern on the DIP switch, then check out your skill at hex
to binary conversion.

19-3. Memory.

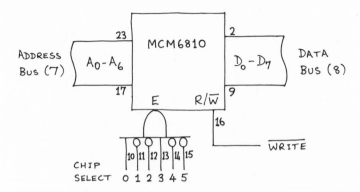

Figure 19.4. Memory pin connections. 24 pin
package. V_{CC} = Pin 24. GND = Pin 1.

The MCM6810 is a 128 location 8 bit wide NMOS memory. It
uses a single +5V supply and has TTL compatible input and
output, so it is very convenient for our use here.

Can you think of a reason for the strange combination of
chip-selects? What must you do with each of them to use
the memory. Connect the READ/WRITE pin to the WRITE
switch used above in part 3 to enable input data to the bus.

When WRITE is low, data are written via the bus into the
memory. Stepping the address counter will cause the same
data to be written into consecutive locations. When WRITE
is high the memory itself asserts the data bus and allows you
to examine the content of each location.

Thoroughly test reading and writing functions with
various bit patterns for data to make sure that no bit is
permanently in the same state, then do the same for several
locations over the full range of addresses.

19-4. Before You Leave.

Do not disassemble all your hard work! The fun hasn't
started yet! Write your name on the breadboard and store it
somewhere safe until next time.

Reading: Chapter 10.12 to end.
                Introduction to this Lab and Z80 Appendix.

Problems: Problems in Text.
                Additional Exercises 7,8,9.

Now comes the moment of truth when you add the Z80
microprocessor to the bus structure you made last time.
There are a few additional details to take care of before we
can run the computer, namely gating to control access to the
bus and to latch the displays (all the $\overline{\text{ENABLE}}$s were tied to
ground before) and provision of a source of clock pulses of
the right specification. These will probably take most of the
session to get running.

   First of all let's introduce the Z80. The pinout is shown in
figure 1. 24 of the 40 pins are taken up with the address (16
bits) and data (8 bits) buses, leaving 16 more for power
connections and control lines. We will at some point use or
at least look at most of these, so it will help to have a
summary of their functions:

$\overline{\text{BUSRQ}}$ Input, active low. When this "bus request" input is
         asserted, the computer releases both the buses
         within a few clock cycles, which then float to the
         open state.   Some control lines are also floated.
         We will connect a switch to this input so that we
         can disconnect the computer from the bus to allow
         us to access the memory directly.

$\overline{\text{BUSAK}}$ Output, active low. The action of $\overline{\text{BUSRQ}}$ is not
         immediate; the computer must complete the
         instruction it is currently executing before freeing
         the bus or else it would not be able the continue
         correct operation afterwards. Therefore, to avoid
         bus contention problems this "bus acknowledge"
         level must be used to arbitrate external access to
         the buses. We will connect the $\overline{\text{ENABLE}}$ of the
         address counter tristate buffer to $\overline{\text{BUSAK}}$ for this
         reason. We will also connect an LED to this to
         indicate when we have manual access to the buses.

$\overline{\text{RESET}}$ Input, active low.   This input initializes the
         computer, clearing all the status flags, and causes
         execution to start at address location 0000.

$\overline{\text{HALT}}$ Output, active low. This output goes low after execution of a 'HALT' instruction and can only return high by the action of $\overline{\text{RESET}}$ or a suitable interrupt. We will run an LED from this to indicate when the computer is running.

$\overline{\text{MREQ}}$ Tristate output, active low. "Memory request" timing signal to indicate when an operation involving memory is active.

$\overline{\text{M1}}$ Tristate output, active low. "Memory cycle 1" timing signal.

$\overline{\text{IORQ}}$ Tristate output, active low. "Input/output request" timing signal indicating that an 'IN' or 'OUT' instruction is being executed. This is used to enable peripheral devices to the data bus.

$\overline{\text{RD}}$ Tristate output, active low. Strobe for "read" from memory or I/O.

$\overline{\text{WR}}$ Tristate output, active low. Strobe for "write" to memory or I/O.

These last five timing control outputs (and the $\overline{\text{RFSH}}$ "dynamic memory refresh", which we do not use) are best understood by looking at the timing diagrams in the Z80 appendix. We will also look at them later today with the oscilloscope. Lastly there are three inputs we will not use at all; each is disabled by tying to +5V:

$\overline{\text{WAIT}}$ Introduces "wait" states to elongate the memory access cycles to allow the use of slower memories.

$\overline{\text{INT}}$ "Interrupt request".

$\overline{\text{NMI}}$ "Non maskable interrupt request" causes a branch to location 0066_{16}. Analogous to 8085 'TRAP'.

20-1. Central Processor.

Connect up your Z80 as shown in figure 20.1. Note that inputs are never left floating in order to avoid erratic operation, but that unused outputs such as the upper address lines are left unconnected. The data bits are in a strange order so remember to double-check their connection to your existing data bus. "$\emptyset$" is the symbol used for the master clock which we build in part 4.

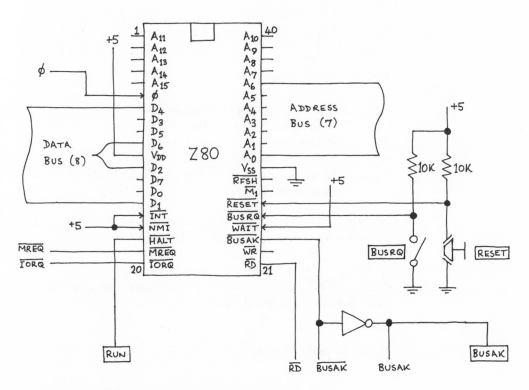

Figure 20.1. Central processor organization.
RUN and BUSAK are both connected to breadboard
LED indicators.

Now connect $\overline{\text{BUSAK}}$ to the two $\overline{\text{ENABLE}}$s of the tristate
buffer that allows the up/down counter to assert the address
bus. This will automatically allow the counter to address the
memory when the computer releases the bus upon
acknowledging your $\overline{\text{BUSRQ}}$ switch.

20-2. Address Display Enable.

It is probably not clear why we cannot leave the address
display enabled permanently. The answer will be more
obvious later when we look at the address bus with the
oscilloscope. In normal operation the Z80 sends out address
information on the bus alternated with "refresh" addresses,
which are used with dynamic memories only (see §11.10).
We are only interested in the former and will be very
confused if we do not screen out the latter from our display.
If you study the timing diagrams in the Z80 appendix you will
see that $\overline{\text{RD}}$ is the correct level needed to achieve this;
logical AND with the clock is needed because of the
extraordinary hold time (50ns) of the latch input to the HP

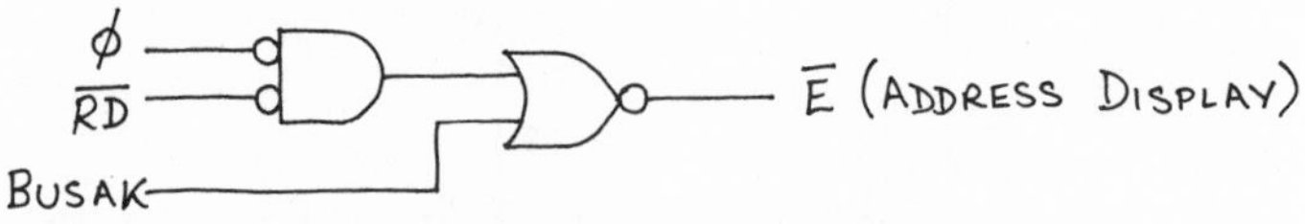

Figure 20.2. Address display enable.
Use a 74LS02 (quad NOR) for the two gates.
ø   is the master CPU clock (Z80 pin 6).
$\overline{RD}$ is the Z80 output pin 21.
BUSAK   is derived from $\overline{BUSAK}$ through the inverter
          in part 20-1.

displays, and logical OR with BUSAK enables the display
when we have manual control of the bus.

When you connect the output of these gates to the
display $\overline{ENABLE}$s remember to **disconnect the ground** that
was connected there before.

20-3. Memory Enable.

$\overline{MREQ}$ ─
$\overline{RD}$ ─ $\overline{E}_{MEM}$
BUSAK─

Figure 20.3. Memory enable gating.
$\overline{MREQ}$, $\overline{RD}$, and BUSAK are all signals you have
available from the CPU in part 1. Use the other half
of the 74LS02 you used in part 2 to build this.

We have to use gates here because the Z80 with its limited
number of pins does not provide a "read from memory"
signal; instead we must use RD AND MREQ. To see why,
consult the timing diagrams in the appendix: MREQ is true
for "memory write" and "refresh" operations as well, which
we must exclude; RD is also true during I/O operations; they
are only both true during "memory read". Logical OR with
BUSAK is needed for the same reason as above in part 2.

What logic would be required to allow the CPU to write
to memory as well? Remember here that the memory R/$\overline{W}$
can no longer be permanently high and that manual writing
to memory must also be accommodated. The Z80 has a $\overline{WR}$
output exactly analogous to $\overline{RD}$. **It is not necessary to
build this modification**; the Z80 has ample registers for all

our calculations, so we never need to write to memory. Of course, if you attempt to write your own programs that require memory writing for this computer, you should make the modification (see appendix), but it can actually be dangerous to have this capability if a program misbehaves and starts overwriting itself!

Disconnect one of the memory $\overline{CS}$ inputs from ground and connect it to $\overline{EMEM}$.

20-4. Clock.

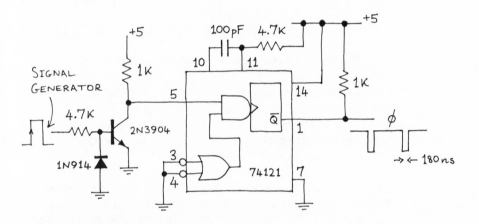

Figure 20.4. Clock circuit.

This circuit has three functions:

1) To produce a negative going pulse which is of short duration at **all** input frequencies. This is the recommended way to clock the Z80.
2) To clean up the waveform to TTL standards (the 74121 has an internal Schmitt trigger on its input).
3) To protect the Z80 against overdriving on its clock input, which might occur if the signal generator were connected directly.

Drive the input from the signal generator and look at the output with the scope. If you have a high frequency input you will be able to see the 200ns pulses at the output.

When you have a valid clock waveform you should test the bus request and acknowledge sequence: Issue of a bus request ($\overline{BUSRQ}$ low, i.e. switch closed) causes the BUSAK LED to come on within a few clock cycles, and its removal causes the LED to go out. Until this sequence works, you cannot proceed. If it fails, recheck the clock with the logic probe and make sure that the $\overline{\text{ENABLE}}$s driven by $\overline{\text{BUSAK}}$ are not still grounded.

When you have successfully obtained a BUSAK, check that you can read from and write to the memory with the address counter and DIP switch, then enter this program.

20-5. Simple Program.

adr	data	label	operation	comments
00	00	BEGIN:	NOP	; "No operation" = do nothing.
01	00		NOP	; Ditto.
02	C3 00 00		JMP BEGIN	; Jump back to first instruction

Program 20.1. "Keep looping". All programs in this book will use this format, which is typical of the output listing of an assembler. Note in particular that the three byte 'JMP' occupies a single line and that there are three entries in the 'data' column. Remember to write all three into successive memory locations when you enter the program.

Program 20.1 causes the computer to loop indefinitely so that we can watch its behavior on the scope. Enter this program into the first five locations of the RAM. Start execution by the sequence: 1) return the bus to the CPU by raising $\overline{\text{BUSRQ}}$ to high, then 2) momentarily bring $\overline{\text{RESET}}$ low, which causes the CPU to fetch the first instruction from location 0000_{16}.

With the clock running at a few Hz, observe the CPU fetch the instructions in sequence. Why does it ever need to look at locations 03 and 04? Then use a higher clock frequency and watch with the scope. What is the highest speed that functions correctly? You might find it useful to look at the $\overline{\text{M1}}$ output (pin 27) which goes low during the first two clock cycles of every instruction (the 'op-code fetch'). You should be able to identify the two NOPs and the JMP in the sequence by their different lengths. If you trigger the scope externally with $\overline{\text{M1}}$ and get a stable trace you can compare the clock on one channel with various control levels on the other. Look at those needed to fill in the timing diagram in figure 20.6. In order to see the floating state of the bus (for the last two lines of the diagram) you should use the trick shown in figure 20.5. Represent the floating state in your diagram with a level halfway between H and L.

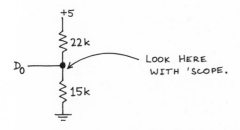

Figure 20.5. Scope sees about 2.0V when the bus is
not asserted. This does not load the bus significantly
otherwise.

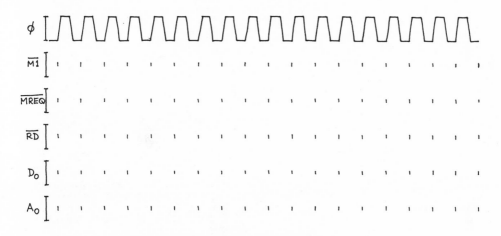

Figure 20.6. Timing diagram. Fill in the required
lines, taking care to maintain the relative phases.

Compare your answer with the timing diagrams in the Z80
appendix and check that your estimation of the lengths of
the instructions agrees with the specified lengths.

You will find the trace for bit 0 of the address bus a
little confusing at first. The repetition period is twice that
of any of the other traces, and corresponds to **two** loops of
program 20.1. What you are seeing is the dynamic memory
"refresh" addresses which the Z80 generates during T_3 and
T_4 of an opcode fetch (see appendix). This address is 7 bits
wide and is incremented every time an opcode fetch is
executed, i.e. every instruction. Here we have 3
instructions in the sequence, so the refresh level on A_0,
which alternates every instruction, takes two passes around
the loop to regain its original phase.

Reading: Chapter 11.1 - 11.4, pp 484-501.
                Read ahead in this lab and make sure you
                understand the programs.

Problems: Problems in text.

Any real microcomputer application will require some form of communication with the outside world. One way to do this is to 'map' input and output devices into memory space so that the computer thinks they are memory. More convenient is the use of the 'IN' and 'OUT' instructions as described in §10.06 and §10.07. We can demonstrate these with a minimum of modification to your circuit, by using the 'data display' circuit as an output device. It is already connected to the data bus, so all we need to do is connect its $\overline{\text{ENABLE}}$ to the CPU's $\overline{\text{IORQ}}$ line (pin 20) and it will display the accumulator every time an 'OUT' is executed. To save you from being overwhelmed by the rate of data output at high clock speeds, we need to separate the 'OUT's with timing loops. This introduces a common trick of microcomputer programming.

21-1. Timing Program.

This technique is frequently used to perform some operation at uniform intervals of time. Rather than use additional hardware to keep time for us we can use a program

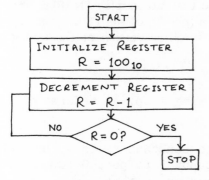

Figure 21.1.   Timing program flow diagram.

structured as in figure 21.1. Program 21.1 is the Z80 coded form. Register B is chosen for counting.

```
00   06 64           MVI  B,100   ; Load register 'B' with 100.
02   05       LOOP:  DCR  B       ; Decrement B:  B = B-1.
03   C2 02 00        JNZ  LOOP    ; Jump back if the result is not
06   76              HLT          ; yet zero.  Otherwise halt.
```

Program 21.1. Timing program. Note the use of
multiple entries in the 'data' (second) column to
indicate an instruction longer than one byte; the
'address' (first) column therefore skips between lines
by one, two or three locations at a time.

Make sure you have an LED connected to the $\overline{HALT}$
output (pin 18) of the CPU, enter the program into memory
and start it with RESET. It stops itself.

The DCR instruction takes 4 machine cycles, the JMP 10.
The loop therefore lasts 100x14xT seconds, where T is the
period of the clock. Verify this with the clock at 100Hz.

Alter the program (not the clock frequency!) to time
exactly 10 seconds and try it out. Think of a way to time
100 seconds with the same clock (you should try this one out
with a 1kHz clock instead).

21-2. Input/Output Programming.

Carry out the modification suggested in the introduction:
Disconnect the ground tied to the data display $\overline{ENABLE}$, and
connect instead the $\overline{IORQ}$ line (pin 20) of the Z80. Then try
out program 21.2 which is a modified version of the timing
program.

```
00   06 64    BEGIN: MVI  B,100   ; Timing loop, as before
02   05       LOOP:  DCR  B
03   C2 02 00        JNZ  LOOP
06   C6 xx           ADI  xx      ; Add constant xx to accumulator
08   D3 00           OUT  0       ; Output A to port 0 (dummy)
0A   C3 00 00        JMP  BEGIN   ; Keep doing this
```

Program 21.2. Display program.

This displays, at suitable intervals, a sequence of hex
numbers, each one greater than the previous one by a preset
constant, xx_{16}. We keep track of the current value in a
second register, the accumulator = register 'A'. Try xx = 03
and use a 1kHz clock.

21-3. Device decoding.

When we have more than one I/O channel we must use some kind of address decoding to distinguish between them. During the execution of an 'IN' or 'OUT' instruction a code called the 'port number' is presented on the address bus while the $\overline{IORQ}$ line is low. Combinational logic could be used to identify the devices, but the most general method is to use a decoder. The 74LS138 is a good choice as it has several enable inputs (two inverted, one not). In the next program we want to be able to read the DIP switch into the accumulator as well as write to the data display, so we must add the following circuit:

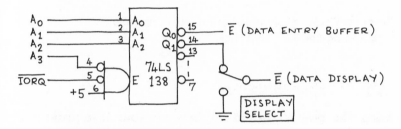

Figure 21.2. Device decoding with a 74LS138 1-to-8 decoder. GND = Pin 8. V_{CC} = Pin 16.

Remember to disconnect the existing circuit from the DIP switch tristate buffer $\overline{ENABLE}$ just before you run the program each time, and restore it before you try to enter data again. The DISPLAY SELECT switch is set to the lower position for entering programs (or otherwise looking at the data bus), and to the upper position to use the display as an output device. You may wish to replace the switch with a gate controlled by BUSAK to do the job automatically; you could do the same for the DIP switch buffer $\overline{ENABLE}$.

The two I/O ports are described in this table:

Port name	I/O	A_7-A_4	A_3	A_2	A_1	A_0	Hex
SWITCH	I	x	0	0	0	0	$x0_{16}$
DISPLA	O	x	0	0	0	1	$x1_{16}$

where 'x' is a don't care state.

Now you can use the switches to set the increment (xx above) while the program is running. The modified version is program 21.3. The "EQU" statements at the top are used by the assembler to define values for the port names we have chosen; there are no data to enter for these lines of code.

```
00  =         SWITCH EQU  0        ; Define port addresses
01  =         DISPLA EQU  1

00  06 64     BEGIN: MVI  B,100    ; As before
02  05        LOOP:  DCR  B
03  C2 02 00         JNZ  LOOP
06  DB 00            IN   SWITCH   ; Read switches into accumulator
08  81              ADD  C        ; Add previous total in reg C
09  00               NOP          ; (see 21-4 below)
0A  4F               MOV  C,A      ; Save new total in C
0B  D3 01            OUT  DISPLA   ; Output new total
0D  C3 00 00         JMP  BEGIN    ; Do it yet again
```

Program 21.3. Display program with increment
supplied on the DIP switches.

21-4. Decimal Arithmetic.

Keep the power turned on to save your last program.
Now change the 'NOP' in address 09 to a 'DAA' (opcode
27_{16}) and run your program again. Impressed?

"Decimal Adjust Accumulator" acts upon the result of
any **arithmetic** operation in the accumulator. It modifies
the answer to what it would have been if the operation had
been performed in 'packed BCD' (i.e. two decimal digits per
byte). This format appears on our displays as a 2-digit
decimal number. To do this, the CPU saves the carry from
bit 3 to bit 4 of every arithmetic operation as flag 'H'; the
DAA looks at H and CY (carry bit from bit 7) as well as the
accumulator to deduce its result. Tricky problem (optional):
Figure out an algorithm to do this.

21-5. Exercise. (Only if you have time)

Write a program to convert the binary number on the
switches into a decimal display. N.B. this is not as trivial as
it may first appear: DAA is **not** a direct binary to decimal
conversion; it only acts on the immediate result of an
arithmetic operation between two **packed BCD** operands.

Hint 1.   Structure your program like this:
          Begin: Read switches
                        Convert to decimal
                        Display
                        Jump to 'Begin'.

Hint 2.   If you run the clock fast enough, you can perform
          the conversion by decrementing the input number
          until zero, while simultaneously counting up (in
          decimal) in a second register. Note that the 'INR'
          instruction does not affect the CY flag so you must
          remember to clear it first (or use an 'ADI   1'
          instead).

Reading: Chapter 11.5 - 11.8, pp 501-517.

Problems: Problems in text.

This exercise, which will take the whole session, is to build a
frequency meter using the microprocessor techniques we
have learned in the last three labs. This demonstrates the
real power of microprocessor-controlled instrumentation,
namely the ability to combine sequential switching operations
with numerical calculation.

   The conventional way to measure frequency is to count
zero crossings of an input signal for a fixed interval of time,
as suggested by figure 22.1. See §14.10 (p618) and figure
14.23 in the text. A microprocessor version of this scheme
would be perfectly practical, however to gain high precision
for low input frequencies, long timing intervals are needed as
indicated. You may have experienced this inconvenience if
you have ever tried using a frequency counter this way.

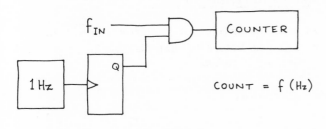

Resolution	Count interval
0.1 Hz	10 sec
0.01 Hz	100 sec etc.

Figure 22.1. Conventional frequency counter block
diagram.

   An alternative approach, which can be implemented when
the ability to do calculations is available, is to measure the
period of the input signal and determine its **reciprocal**,
using the scheme of figure 22.2. See §14.10 (figure 14.25,
p619) in the text. This is what we will do today. The
design we will use gives us a range of 2.5 to 9.9Hz full scale
(two decimal digit display); a simple rearrangement gives
0.99Hz full scale with 0.01Hz resolution.

   We will first construct the hardware for a computer
comtrolled realization of this scheme, then develop the

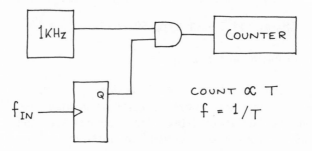

Figure 22.2. Period measurement scheme.

software to make a frequency meter. We need a total of four I/O devices:

    1) Frequency input sensor.
    2) Counter control (reset).
    3) Counter read.
    4) Display, preferably decimal.

22-1. Modifications to Existing Circuit.

1)   Reconnect the DIP switch tristate buffer $\overline{\text{ENABLE}}$ to the WRITE switch as it was before lab exercise 21-3. Leave the 74LS138; you will need it today.

2)   New clock generator.

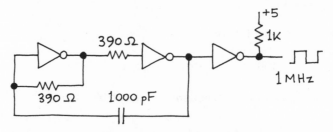

Figure 22.3. TTL 1MHz clock generator.
Important: Use a 7404, not a 74LS04 here.

Unless you happen to have two signal generators available, you will need to use yours as a variable frequency source. Though not an orthodox TTL application, this circuit has never been known not to work.

22-2. Counter Hardware.

We use the breadboard 1kHz clock as a frequency standard; any deviation from the exact value can be compensated

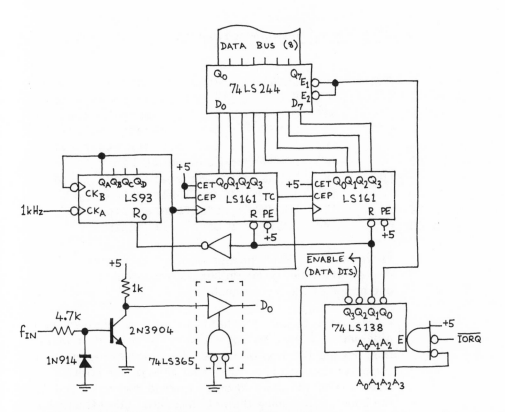

Figure 22.4. Counter hardware. You will need to
refer again to the pinout appendix. Note the unusual
choice of power supply pins with the 74LS93. You
may substitute practically any tristate buffer for the
74LS365.

later. A precision version of this instrument would use a
crystal oscillator here. The correct scaling for a 9.9Hz
meter range is to make the least significant bit of the count
correspond to a time interval of 2ms (see the calculation
below), so we use a divide-by-two between the clock and
the 8-bit timer which is read onto the data bus. For a
0.99Hz range, we will need divide-by-16 in the prescaler;
make sure you know how to arrange that when the time
comes.

In a fancy version of the instrument we might include a
variable gain DC amplifier on the signal input, with an
adjustable threshold level, so that we could accommodate all
kinds of input signals. For simplicity here we will use a large
amplitude square wave from the signal generator as input.
When you power up the circuit check that the signal is
reaching the tristate buffer, using your logic probe.

22-3. Period Measurment Program.

```
00   DB ww      BEGIN: IN    COUNT    ; Read current count value
02   D3 xx             OUT   RESET    ; Reset pulse to counter
04   D3 yy             OUT   DISPLA   ; Display current count
06   DB zz      LOOP1: IN    SENSE    ; Get comparator as bit 0
08   E6 01             ANI   1        ; Mask off all other bits
0A   C2 06 00          JNZ   LOOP1    ; Loop until bit = 0
0D   DB zz      LOOP2: IN    SENSE    ; Same again: keep looping
0F   E6 01             ANI   1        ; until bit = 1
11   CA 0D 00          JZ    LOOP2    ; ("software flip-flop")
14   C3 00 00          JMP   BEGIN    ; Do it all again
```

Program 22.1. Period measurement. This simply
displays the timer count (in hex) corresponding to
one period of the input signal.

We want to test the hardware with a relatively simple
program. Program 22.1 does this. The method of realizing
the toggling flip flop of figure 22.2 deserves mention. The
computer needs to know the exact time of the positive (say)
transition of the input in order to read the count and reset
the counter for timing the next interval. One effective way
to achieve this is to connect the input (once brought to TTL
levels) to the $\overline{NMI}$ ("non maskable interrupt") pin of the
CPU; the entire program would then be an interrupt service
routine ending with a 'HLT'. Instead, we choose to do this
with a 'software flip flop' which says "wait until $D_0 = 0$,
then wait until $D_0 = 1$, then service the counter", thereby
reading the count shortly after the rising edge of the input.
The loop samples the state of the signal every 29 clock
cycles. There is therefore a 29 microsecond uncertainty in
the time of the transition (1MHz clock). There is in addition
a delay of about 10 microseconds between reading the
counter and resetting it. Both of these errors contribute to
the inaccuracy of the measured period, but both are
insignificant compared with the resolution of the counter,
which is 2 **milli**seconds. Question: Why does the delay
between detecting the rising edge and reading the count **not**
contribute error to the period measurement?

Make yourself a table of the port assigments in the
manner of section 21-3 to fill in the values of the device
codes ww, xx, yy, and zz in the assembly listing for the port
names 'COUNT', 'RESET', 'DISPLA' and 'SENSE'. Derive
these from figure 22.4 or from your own circuit if you have
assigned them differently. Enter the program and check out

your circuit. Find and write down for future reference the range of input frequencies for which the count is representative of the period. Why is this range limited at each end?

22-4. Decimal Frequency Readout.

```
00  =                COUNT  EQU  00H       ; Define ports
01  =                RESET  EQU  01H
02  =                DISPLA EQU  02H
03  =                SENSE  EQU  03H

00  DB 00     BEGIN:  IN     COUNT    ; Read current count value
02  D3 01             OUT    RESET    ; Reset pulse to counter
04  06 FF             MVI    B,0FFH
06  2F                CMA             ; Set up register pair BC with
07  3C                INR    A        ; 2's complement of count
08  4F                MOV    C,A
09  21 88 13          LXI    H,1388H  ; Load numerator into HL
0C  AF                XRA    A        ; Clear A.  Start division:
0D  B7        DIV:    ORA    A        ; Clear CY for correct DAA
0E  3C                INR    A
0F  27                DAA             ; Decimal count up A
10  09                DAD    B        ; 16 bit add BC (-count) to HL
11  DA 0D 00          JC     DIV      ; Loop until division done
14  D3 02             OUT    DISPLA   ; Display result
16  DB 03     LOOP1:  IN     SENSE    ; Get comparator as bit 0
18  E6 01             ANI    1        ; Mask off all other bits
1A  C2 16 00          JNZ    LOOP1    ; Loop until bit = 0
1D  DB 03     LOOP2:  IN     SENSE    ; Same again: keep looping
1F  E6 01             ANI    1        ; until bit = 1
21  CA 1D 00          JZ     LOOP2    ; ("software flip-flop")
24  C3 00 00          JMP    BEGIN    ; Do it again
```

Program 22.2. Frequency meter program. This is a modified version of program 22.1, with a routine inserted to calculate the reciprocal of the count in decimal. Note that double register (16 bit) arithmetic is used (see §11.3, p491).

To turn the period into a numerical frequency we must calculate its reciprocal. This is done in program 22.2. The algorithm is crude: "count the number of times the denominator will subtract from (i.e. its 2's complement will add to) the numerator before an overflow occurs". Note that the carry ('CY') is **set** every time a successful subtraction occurs and only left clear when one too many subtractions has been done. If you don't understand why, see §8.03 (p320) about 2's complement arithmetic. No attention is paid to rounding the quotient to the nearest integer; it is rounded down, thus contributing an extra half digit of error (on average). You might think of a way to correct this: It requires a few extra lines of code.

The numerator value of 5000_{10} is chosen so that a frequency of f Hz will produce a display of "10f", which you read as n.m Hz where n and m are the two digits. The clock frequency actually going into the counter is half of the breadboard 1kHz clock, i.e. 500Hz (see figure 22.4). In one period (1/f seconds) 500/f counts are accumulated and "10f" must be displayed; therefore,

$$\text{Numerator} = 10f \times 500/f = 5000.$$

What is the lowest frequency that you can measure with this sceme? How can you increase the range?

First of all, check the meter for linearity; you will probably find that the absolute calibration is inaccurate, but the linearity should be good. Why does the display sometimes skip two values at a time, e.g. between 8.4 and 8.6Hz? Finally calibrate your meter properly by accurately measuring the 1kHz breadboard clock with the lab frequency counter, then recalculate the value of the numerator using the expression above, and test the accuracy of the finished instrument.

22-6. 0.99Hz Range.

Reconfigure your circuit to use the full divide-by-16 prescaler between the 1kHz clock and the timing counter (move the tap on the 74LS93 in figure 22.4). Calculate the new value of numerator required to perform the division correctly and try it. If you want to impress yourself with the advantage this frequency measuring technique has over the conventional counter (and you have lots of patience), hook up the lab counter alongside and compare!

Reading: Chapter 11 to end.

Problems: Problems in text.

Today we will finish off the series of microcomputer
examples with a look at analog-digital techniques. The basic
principle is the same as before, that all the hardware control
signals are connected to different I/O ports, so that
sequential operations become the sequential 'IN's and 'OUT's
of a control program. The main project this time will be to
construct a digital voltmeter with decimal readout, based on
the popular (and cheap) 1408 8-bit binary D/A converter
that we met in lab 17; the computer will therefore have to
do a binary to decimal conversion. We will finish up with an
example of computer graphics, a simple project which always
gives a lot of satisfaction.

23-1. D/A Conversion.

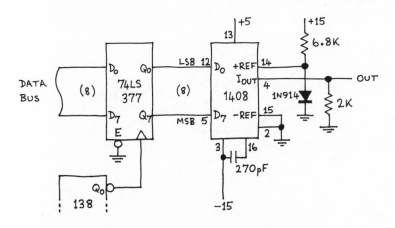

Figure 23.1. 1408 D/A converter attatched to the
data bus through a 74LS377 octal latch.
The 74LS138 decoder is already part of your existing
circuit; device code 0 is used for the D/A output
(device name 'LATCH').

It is necessary to hold the data sent out on the bus during
the conversion, which explains the need for a latch in figure
23.1. The 74LS377 is a positive edge triggered 8-bit latch
which loads from the data bus on the rising edge of the $\overline{\text{IORQ}}$
pulse, a time when the data lines are guaranteed to be stable
(see the I/O timing diagram in the Z80 appendix). Check

out the D/A action with a simple program:

```
00  =            LATCH EQU   0       ; Define port 'LATCH'

00  D3 00        BEGIN: OUT  LATCH   ; Latch contents of A to convert
02  3C                  INR  A       ; Increment A
03  C3 00 00            JMP  BEGIN   ; Repeat
```

Program 23.1. D/A test procedure.

If you have not done so already, return to the original (monostable) clock circuit of lab 20 (figure 20.4), so that you can vary the speed.   Then check that you get a satisfactory 4 volt ramp voltage at the output; use the voltmeter (slow clock) or oscilloscope (fast clock). What is the measured settling time of the D/A when used in this way? If the device using the analog output requires that it must have settled by the beginning of the next instruction (say after 10 microseconds), does this circuit meet the requirement? If it does not, can you think of a 'soft' way around the problem?

23-2. Tracking A/D Converter.

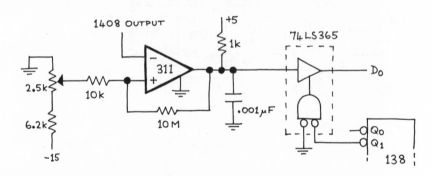

Figure 23.2. Comparator input.
Device name 'COMPAR', device code 1.
The voltage input is just a potentiometer to -15V.

Retain the use of device code 2 for 'DISPLA', hence the connection from $\overline{Q_2}$ of the decoder to $\overline{ENABLE}$ (data display). Figure 23.3 shows a flow diagram of the procedure for analog to digital conversion with this hardware, which has exactly the same function as the tracking converter we built in lab 17. Watch its behavior, which should be familiar to you from before, as you run the 2.5k potentiometer up and down. Try it with both fast and slow clocks to watch the tracking closely.

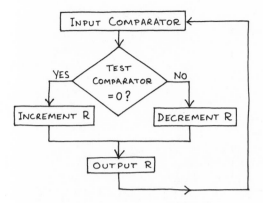

Figure 23.3. Tracking A/D flow diagram.
This corresponds to program 23.2.

```
00  =            LATCH  EQU  0       ; Define I/O ports
01  =            COMPAR EQU  1
02  =            DISPLA EQU  2

00  DB 01        BEGIN: IN    COMPAR ; Test comparator. Sets bit 0
02  E6 01               ANI   1      ; Select bit 0.  Set flags
04  CA 10 00            JZ    CTUP   ; Branch to increment,
07  0D                  DCR   C      ; otherwise decrement reg C
08  79           BACK:  MOV   A,C    ; Move new value to A
09  D3 00               OUT   LATCH  ; Send to DAC
0B  D3 02               OUT   DISPLA ; Display new value
0D  C3 00 00            JMP   BEGIN  ; Repeat
10  0C           CTUP:  INR   C      ; Increment C
11  C3 08 00            JMP   BACK   ; Return to main loop
```

Program 23.2. Tracking A/D converter.

3-3. Digital Voltmeter.

The distinctions between simple A/D converters and digital
voltmeters are that, because the latter must interface with
yourself, the user, the display should be:

1) in volts (rather than 'arbitrary units'),
2) in decimal,
3) changing at most maybe 10 times a second.

The last constraint is to avoid the effect of seeing two
adjacent digits superimposed. It is not sufficient to slow
down the clock rate to avoid this problem, as it would then
take the meter many seconds to settle to its final value.

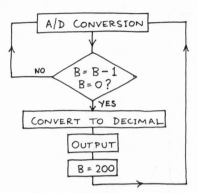

Figure 23.4. Digital voltmeter flow chart.

```
00  =           LATCH  EQU  0      ; Define I/O ports
01  =           COMPAR EQU  1
02  =           DISPLA EQU  2

00  DB 01       BEGIN: IN   COMPAR ; Test comparator. Sets bit 0
02  E6 01              ANI  1      ; Select bit 0
04  EE 01              XRI  1      ; Complement bit 0
06  07                 RLC         ; Rotate to bit 1
07  3D                 DCR  A      ; Decrement A: A = +/-1 now
08  81                 ADD  C      ; Add previous value in C
09  D3 00              OUT  LATCH  ; Send new value to DAC
0B  4F                 MOV  C,A    ; and save it in C again
0C  05                 DCR  B      ; Decrement count register B
0D  C2 00 00           JNZ  BEGIN  ; and loop until count finished
10  B7                 ORA  A      ; Clear CY flag before rotate
11  1F                 RAR         ; Shift right. Waste bit 0
12  3C                 INR  A      ; Add 1: avoids error for 0
13  57                 MOV  D,A    ; Move answer to D
14  AF                 XRA  A      ; Clear A.
15  3C          LOOP:  INR  A      ; Start decimal conversion:
16  27                 DAA         ; Decimal count up A,
17  15                 DCR  D      ; while count down D
18  C2 15 00           JNZ  LOOP   ; until all gone.
1B  3D                 DCR  A      ; Remove extra count
1C  27                 DAA         ; (in decimal)
1D  D3 02              OUT  DISPLA ; Display new value
1F  06 C8              MVI  B,200  ; Reload count register
21  C3 00 00           JMP  BEGIN  ; Repeat
```

Program 23.3. Digital voltmeter.

Instead, we must arrange to change the display much less often than updating the D/A latch. Figure 23.4 shows the strategy and program 23.3 is a way of achieving this. The maximum decimal display value obtainable is 99_{10}, less than half the precision of the A/D, which is 256 increments. Since we only need seven binary bits to get full scale, and the result is more accurate if we discard the LSB than the MSB, we rotate the accumulator right once (address 0011_{16}) before converting to decimal.

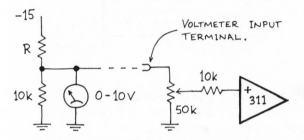

Figure 23.5. Calibration and test circuit.

We can now use the potentiometer at the input for calibration as in figure 23.5, and use a resistance substitution box in series with the final voltmeter's internal resistance for testing purposes. Instead, you might like to use a very low frequency AC voltage at the input.

23-4. X-Y Display.

We can now use two DACs to put out both the X and Y coordinates of a point on the scope screen. To avoid the need for any brightness information on the scope, which would be necessary to blank out the transient state between moving X and moving Y to their respective ports, we can use the simple trick of figure 23.6. X and Y are sent simultaneously to the **same** port as the upper and lower halves of one byte. The limitation is in resolution: The addressable space is a 16x16 array of dots.

Try it first of all with program 21.2 which should display diagonal lines of dots. How is the slope of the lines determined? Lastly, try program 23.4 which displays in sequence a list of coordinates stored in a table in memory. Design on squared paper a pattern of dots (your initials perhaps, or your views of this book) for your coordinates.

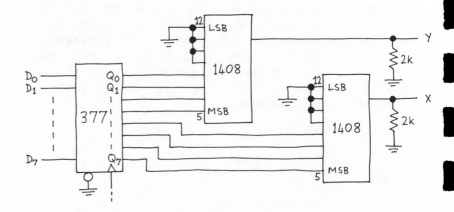

Figure 23.6. X-Y display. If you use the **most** significant four bits of each 1408 as input, the overall swing of the output will be greater and so the relative level of noise will be lower. You will get cleaner looking spots that way.

```
00   21 10 00   BEGIN: LXI   H,TABLE  ; Load HL with 1st addr of table
03   06 xx             MVI   B,xx     ; Load count: xx = table length
05   7E         LOOP:  MOV   A,M      ; Get coords from table to acc.
06   D3 00             OUT   0        ; Output to scope (dummy port)
08   23               INX   H        ; Increment HL (double reg)
09   05               DCR   B        ; Decrement count in B,
0A   C2 05 00          JNZ   LOOP     ; and loop until zero
0D   C3 00 00          JMP   BEGIN    ; And repeat
10              TABLE: 0             ; Here begins your coord table..
```

Program 23.4. List display program. The table of coordinates begins at location 0010_{16}, and its length (i.e. number of points) goes in location 0004_{16}.

The following lists comprise a complete inventory of the small parts required for these laboratory exercises. See the Introduction for the major pieces of equipment needed.

Unless stated otherwise, linear IC's should be purchased in miniDIP packages, and digital IC's in plastic DIP's; both should be the inexpensive "commercial" temperature range (0°C - 70°C). For the linear IC's where package styles are specified, see the data sheets for the appropriate suffixes, which vary from one manufacturer to another. We have indicated in **boldface** those parts that are used most frequently, and should be stocked in quantity. Small resistor assortments, available from Ohmite or Stackpole in convenient cabinets, are a good way to provide the 1/4 watt resistors. If the potentiometers are not provided as part of the breadboards, the cermet single-turn TO-5 pattern PC-mounting types (e.g., Bourns 3386P, CTS 362Y) are particularly convenient.

It is best to solder short lengths of 22-gauge bare wire to the bulky components (toggle and push-button switches, lamps, inductors), and a pair of insulated 22-gauge solid leads to the filament transformer and loudspeaker, for easy insertion into the prototyping boards.

Resistors		Capacitors
22 ohms	6.8k	33pF mica or disc
27	**7.5k**	50pF "
33	8.2k	100pF "
47	**10k**	270pF "
100	11k	330pF "
120	12k	560pF "
150	15k	680pF "
180	20k	0.001uF mylar
220	22k	0.0033uF "
240	33k	**0.01uF** "
270	47k	**0.1uF** "
390	56k	1uF, 35V tantalum
470	68k	**4.7uF, 50V** "
680	82k	15uF, 20V "
750	**100k**	68uF, 20V "
1k	130k	500uF, 25V electrolytic
1.5k	150k	
1.6k	200k	
2k	330k	
2.2k	390k	
2.7k	470k	
3.3k	**1M**	
3.9k	4.3M	
4.7k	4.7M	
5.6k	10M	
6.2k		

Diodes
1N749
1N914
1N4004

Transistors
2N3565
2N3725
2N3904
2N3906
2N3958
2N5485
2N5962

Linear IC's
311
317 (TO-39)
355
358
555
566
723 (DIP)
741
MC1408L8P
CA3096
78L05 (TO-92)

Controls and Switches
2.5k pot
10k pot
50k pot
SPDT switch
SPDT push button
8-bit DIP switch

TTL IC's
74LS00
74LS02
74LS04
7404
74LS74
74LS83
74LS85
74LS86
74LS90
74LS93
74LS107
74121
74LS125
74LS138
74LS151
74LS161
74LS164
74LS191
74LS193
74LS244
74LS365
74LS377

CMOS IC's
4011
4020
4046
4066
4069UB

Miscellaneous IC's
Z80 CPU
MCM6810
HP 5082-7340

Miscellaneous
red LED
#47 lamp
#1869 (or #344) lamp
7mH inductor (UTC MQE-1)
6.3VCT filament transformer
small loudspeaker

The quantity of reference data needed for a microprocessor is so great it takes up an entire appendix to itself.

The reason for choosing the Z80 as the basis for the series of microprocessor labs is that, because its registers are static, it can be run at very low clock speeds. This is essential for debugging when a bare computer, without any monitor program, is used. In order to remain consistent with the text, we choose to use 8085 mnemonics for the instructions and restrict ourselves to that 78 instruction subset of the 158 instructions in the full Z80 set (as is the practice of many Z80 software manufacturers to ensure compatibility). The larger instruction set includes the additional move and arithmetic operations that apply to the new registers (the Z80 has 11 more registers than the 8085: auxilliary A-L plus index registers) as well as new operations such as automatic loop counting and block move features. Most of the new instructions have 2 byte opcodes, since 245 of the 256 possible 1 byte codes are utilized in the 8085 instruction set. Refer to the manufacturer's data sheets for further information if you want to experiment with these.

If you attempt to write your own programs for the computer described in this book, use the opcode table at the end of this appendix to hand-assemble your code. The number of machine cycles for each instruction is also given; these are only slightly different from those of the 8085. All other information is available in chapter 11 of the text. The way to determine which flags are set by any instruction is to look at table 11.2, for instance.

Remember though if you try your own programs that the computer you build in labs 19 and 20 is **unable to write to memory.** This not only prevents you from using 'MOV M,r', 'STA' etc, but also from using the stack operations, including subroutine calls and the interrupt system, and so it can be very restrictive. The way to allow the computer to write to memory is to modify the circuit of figure 20.3 to that of figure B.1, which now drives the memory R/$\overline{\text{W}}$ as well as $\overline{\text{EMEM}}$.

B-1. Timing Diagrams.

The mimimum number of clock cycles taken by an instruction is four. This is the amount of time required to fetch the

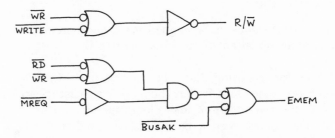

Figure B.1. Memory enable gating to allow the computer to write directly. EMEM goes to one of the (positive true) memory CHIP SELECTs, not to a $\overline{\text{CHIP SELECT}}$ as it did before. $\overline{\text{WRITE}}$ is the connection to the manual WRITE switch. Note that it is now $\overline{\text{BUSAK}}$ that is used, not BUSAK.

opcode from memory and decode it. If no more memory references are needed, such as to move data from one register to another or to perform register arithmetic, the execution is complete at this point; if **sequential** register operations are needed, this may be stretched by a few periods (e.g. 'INX' needs 6 periods, 'DAD' needs 11). Thereafter, each memory access requires an additional three periods, so an 'ADD M' which must read from memory lasts for 4+3 = 7 periods.

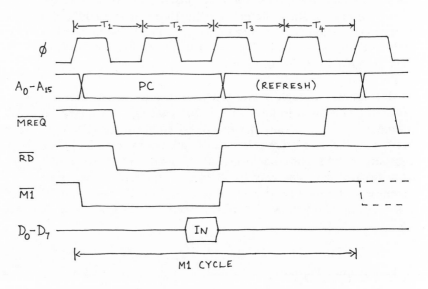

Figure B.2. Instruction opcode fetch ('M1' cycle) timing.

The address of the required instruction in the program
counter ("PC") is placed on the address bus first of all at the
beginning of T_1. A half clock period later, when the bus is
stable, $\overline{MREQ}$ goes low together with $\overline{RD}$ which enable the
memory for reading. The memory then has one and a half
clock periods to produce the data which are latched at the
beginning of T_3. During T_3 and T_4 the address bus is used
for a dynamic memory refresh address (7 bits only) and $\overline{RFSH}$
goes low to enable this; meanwhile the CPU is decoding the
instruction it has received. The output $\overline{M1}$ is active during
the first half of an opcode fetch and only at this time, so can
be used to identify the beginning of each instruction.

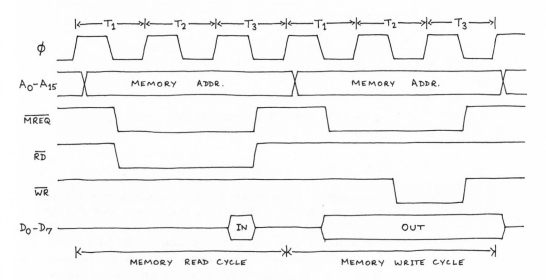

Figure B.3. Memory read and write cycles.

The use of the $\overline{MREQ}$, $\overline{RD}$ lines and the buses during a
memory read is similar to that of the opcode fetch. During
a memory write, the data are presented along with the $\overline{MREQ}$
enable, but the $\overline{WR}$ strobe is witheld to allow for the setup
("access") time of the memory.

The input and output waveform is identical to the
memory read and write, except that an additional T_2 or
"wait" state is generated by the CPU so that longer setup
times for the I/O devices can be accommodated and $\overline{IORQ}$
becomes active instead of $\overline{MREQ}$.

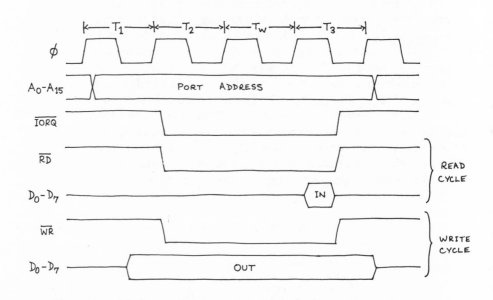

Figure B.4. Input and output cycles.

B-2. Table of Opcodes.

Now follows a complete table of the opcodes for the 78
instructions of the 8085 with a brief functional description.
Note that bits are missing from certain codes for a **register**
or **condition** operand which should be filled in from the
corresponding table in the "definition" section below. The
number of extension bytes of data to complete the
instruction is indicated by the number of 'd's in the operand.

Move, Load and Store		Opcode	Cycles
MOV r,r'	Move register' to register	0 1 r r r r'r'r'	4 [7]
MVI r,d	Move immediate to register	0 0 r r r 1 1 0	7 [10]
LXI rp,dd	Load immed. to reg. pair	0 0 p p 0 0 0 1	10
STAX B	Store A indirect by BC	0 0 0 0 0 0 1 0	7
STAX D	Store A indirect by DE	0 0 0 1 0 0 1 0	7
LDAX B	Load A indirect by BC	0 0 0 0 1 0 1 0	7
LDAX D	Load A indirect by DE	0 0 0 1 1 0 1 0	7
STA dd	Store A direct	0 0 1 1 0 0 1 0	13
LDA dd	Load A direct	0 0 1 1 1 0 1 0	13
SHLD dd	Store H,L direct	0 0 1 0 0 0 1 0	16
LHLD dd	Load H,L direct	0 0 1 0 1 0 1 0	16
XCHG	Exchange DE and HL	1 1 1 0 1 0 1 1	4

Increment and Decrement		Opcode	Cycles
INR r	Increment register	0 0 r r r 1 0 0	4 [11]
DCR r	Decrement register	0 0 r r r 1 0 1	4 [11]
INX rp	Increment register pair	0 0 p p 0 0 1 1	6
DCX rp	Decrement register pair	0 0 p p 1 0 1 1	6

Arithmetic and Logical			
ADD r	Add register to A	1 0 0 0 0 r r r	4 [7]
ADC r	Add reg. to A with carry	1 0 0 0 1 r r r	4 [7]
SUB r	Subtract register from A	1 0 0 1 0 r r r	4 [7]
SBB r	Subtract with borrow	1 0 0 1 1 r r r	4 [7]
ANA r	AND register with A	1 0 1 0 0 r r r	4 [7]
XRA r	Exclusive OR reg with A	1 0 1 0 1 r r r	4 [7]
ORA r	OR register with A	1 0 1 1 0 r r r	4 [7]
CMA r	Compare register with A	1 0 1 1 1 r r r	4 [7]
ADI d	Add immediate to A	1 1 0 0 0 1 1 0	7
ACI d	Add immed. with carry	1 1 0 0 1 1 1 0	7
SUI d	Subtract immed. from A	1 1 0 1 0 1 1 0	7
SBI d	Subtract immed. w. borrow	1 1 0 1 1 1 1 0	7
ANI d	AND immediate with A	1 1 1 0 0 1 1 0	7
XRI d	Exclusive OR immed. w. A	1 1 1 0 1 1 1 0	7
ORI d	OR immediate with A	1 1 1 1 0 1 1 0	7
CPI d	Compare immed. with A	1 1 1 1 1 1 1 0	7
DAD rp	Add reg. pair to HL	0 0 p p 1 0 0 1	11

Accumulator and Flag Operations			
RLC	Rotate A left	0 0 0 0 0 1 1 1	4
RRC	Rotate A right	0 0 0 0 1 1 1 1	4
RAL	Rotate A left via carry	0 0 0 1 0 1 1 1	4
RAR	Rotate A right via carry	0 0 0 1 1 1 1 1	4
DAA	Decimal adjust accumulator	0 0 1 0 0 1 1 1	4
CMA	Complement accumulator	0 0 1 0 1 1 1 1	4
STC	Set carry	0 0 1 1 0 1 1 1	4
CMC	Complement carry	0 0 1 1 1 1 1 1	4

I/O, Control and Stack Operations			
IN d	Input (port addr. = d)	1 1 0 1 1 0 1 1	10
OUT d	Output (port addr. = d)	1 1 0 1 0 0 1 1	11
EI	Enable interrupts	1 1 1 1 1 0 1 1	4
DI	Disable interrupts	1 1 1 1 0 0 1 1	4
NOP	No operation	0 0 0 0 0 0 0 0	4
HLT	Halt	0 1 1 1 0 1 1 0	4
PUSH rp	Push reg. pair on stack	1 1 p p 0 1 0 1	11
POP rp	Pop reg. pair from stack	1 1 p p 0 0 0 1	10
XTHL	Exchange HL w top of stack	1 1 1 0 0 0 1 1	19
SPHL	Move HL to SP	1 1 1 1 1 0 0 1	6

Transfer of Control   Opcode   Cycles

			Opcode	Cycles
JMP dd	Jump unconditional	1 1 0 0 0 0 1 1	10
Jcc dd	Jump on condition cc	1 1 c c c 0 1 0	10
CALL dd	Call unconditional	1 1 0 0 1 1 0 1	17
Ccc dd	Call on condition cc	1 1 c c c 1 0 0	17 (10)
RET	Return from call	1 1 0 0 1 0 0 1	10
Rcc	Return on condition cc	1 1 c c c 0 0 0	11 (5)
RST n	Restart at location 8*n	1 1 n n n 1 1 1	11
PCHL	Move HL to PC	1 1 1 0 1 0 0 1	4

Definitions

1) Data fields

'd' One byte immediate data.   Instr. length = 2 bytes.
'dd' Two byte address.   Instr. length = 3 bytes.
     All other instructions are one byte long.

2) Register fields

'r'	r r r		'rp'	p p	
B	0 0 0		BC	0 0	
C	0 0 1		DE	0 1	
D	0 1 0		HL	1 0	
E	0 1 1		SP	1 1	[not stack ops.]
H	1 0 0		PSW	1 1	[stack ops.]
L	1 0 1				
M	1 1 0	['M' = (HL)]			
A	1 1 1				

3) Condition codes

'cc'	c c c	Condition
NZ	0 0 0	Non zero
Z	0 0 1	Zero
NC	0 1 0	No carry
C	0 1 1	Carry
PO	1 0 0	Parity odd
PE	1 0 1	Parity even
P	1 1 0	Positive
M	1 1 1	Minus

Example: 'JC' (opcode 1101 1010) = "Jump on carry"

4) Number of clock cycles

N   Number of cycles to complete instruction.
[N] Number of cycles when r = M (memory access).
(N) Number of cycles if condition is not met.

TRANSISTORS

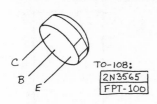

TO-108:
2N3565
FPT-100

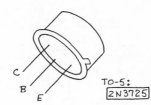

TO-5:
2N3725

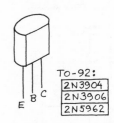

TO-92:
2N3904
2N3906
2N5962

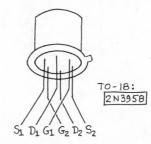

TO-18:
2N3958

S_1 D_1 G_1 G_2 D_2 S_2

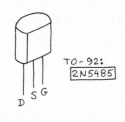

TO-92:
2N5485

D S G

LINEAR IC's

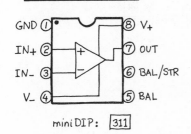

GND ① — ⑧ V+
IN+ ② — ⑦ OUT
IN− ③ — ⑥ BAL/STR
V− ④ — ⑤ BAL

miniDIP: 311

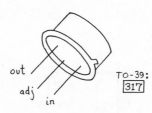

out
adj
in

TO-39:
317

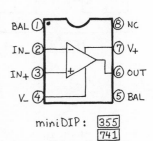

BAL ① — ⑧ NC
IN− ② — ⑦ V+
IN+ ③ — ⑥ OUT
V− ④ — ⑤ BAL

miniDIP: 355
741

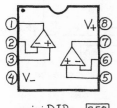

① — V+ ⑧
② — ⑦
③ — ⑥
④ V− — ⑤

miniDIP: 358

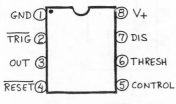

GND ① — ⑧ V+
TRIG ② — ⑦ DIS
OUT ③ — ⑥ THRESH
RESET ④ — ⑤ CONTROL

miniDIP: 555

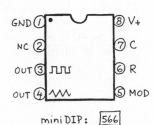

GND ① — ⑧ V+
NC ② — ⑦ C
OUT ③ — ⑥ R
OUT ④ — ⑤ MOD

miniDIP: 566

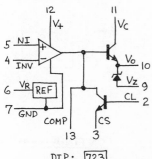

DIP: 723

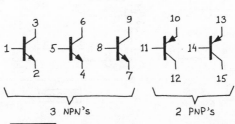

3 NPN's   2 PNP's
CA3096 16-pin DIP (substrate = pin 16)

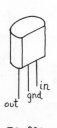

out gnd in

TO-92:
78L05

TTL IC's

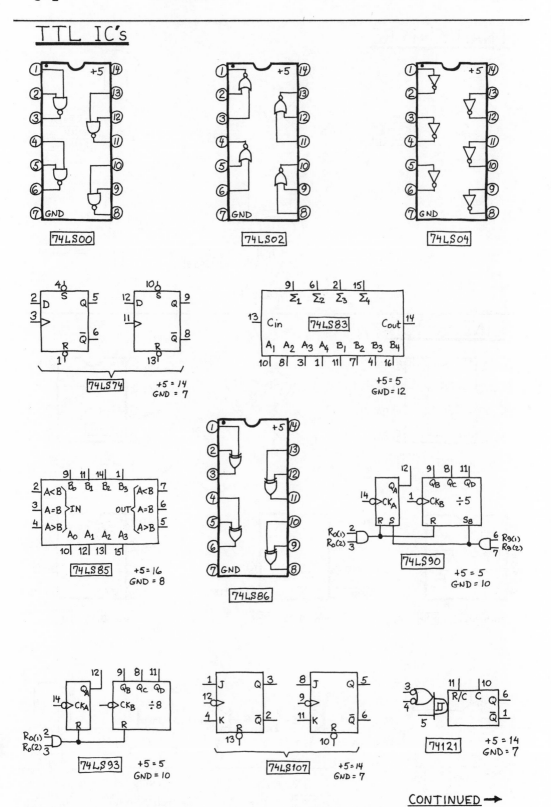

74LS00

74LS02

74LS04

74LS74   +5 = 14
GND = 7

74LS83   +5 = 5
GND = 12

74LS85   +5 = 16
GND = 8

74LS86

74LS90   +5 = 5
GND = 10

74LS93   +5 = 5
GND = 10

74LS107   +5 = 14
GND = 7

74121   +5 = 14
GND = 7

CONTINUED →

TTL IC'S (CONT'D)

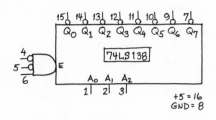

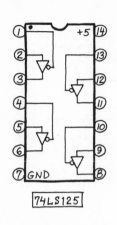

74LS125

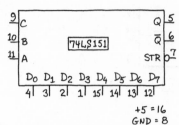

+5 = 16
GND = 8

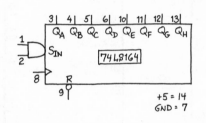

+5 = 16
GND = 8

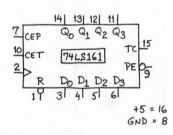

+5 = 16
GND = 8

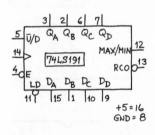

+5 = 14
GND = 7

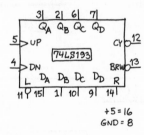

+5 = 16
GND = 8

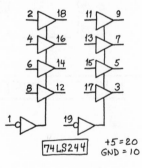

74LS244

+5 = 20
GND = 10

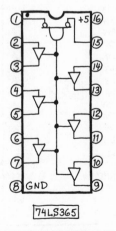

74LS365

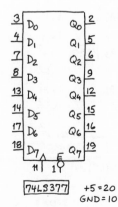

74LS377

+5 = 20
GND = 10

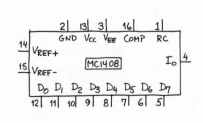

+5 = 16 values noted in diagrams

CMOS IC's

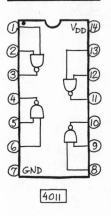

4011

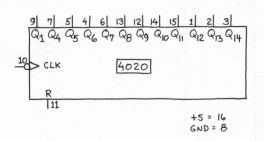

4020

+5 = 16
GND = 8

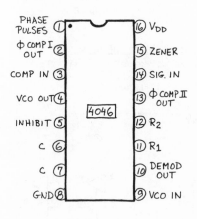

4046

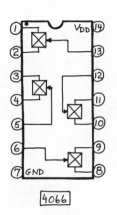

4066

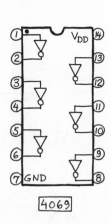

4069

MISCELLANEOUS IC's

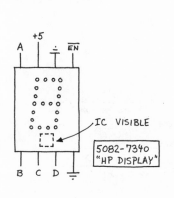

5082-7340
"HP DISPLAY"

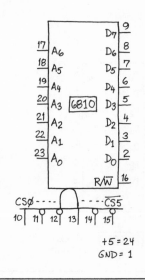

6810

+5 = 24
GND = 1

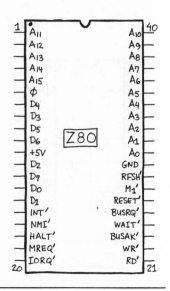

Z80